GOAT FARMING

Alan Mowlem

FARMING PRESS

First published 1988

Copyright © Alan Mowlem, 1988

British Library Cataloguing in Publication Data

Mowlem, Alan
    Goat farming.
    1. Livestock : Goats. Production
    I. Title
    636.3'9

    ISBN 0-85236-183-1

Cover photographs by Holt Studios,
Hungerford, Berks RG17 0NF

Published by Farming Press Books
4 Friars Courtyard, 30–32 Princes Street
Ipswich IP1 1RJ, United Kingdom

*Distributed in North America
by Diamond Farm Enterprises,
Box 537, Alexandria Bay, NY 13607, USA*

Phototypeset by Galleon Photosetting, Ipswich
Printed and bound in Great Britain by
Redwood Burn Limited, Trowbridge, Wiltshire

Contents

Preface

The idea of farming goats has taken a long time to become accepted in the Western world and in Britain in particular. This can mainly be attributed to the unfortunate stigma attached to goats and goatkeepers. The reasons for this lay mainly with the animal which, although it has served man almost longer than any other species, still maintains an independence of character more usually associated with wild animals. This independent character results in behaviour patterns quite unlike those of any other farm livestock and of a kind to drive some stockmen, not used to goats, to distraction! It is fair to say that the goat seems to tolerate domestication only because it suits it—a situation that can be compared with the domestic cat.

If one ignores the character and behaviour of the goat and just considers its performance, it is impressive. It has the ability to survive in areas of the world that will not support other livestock and in such environments it will produce products of value to man. Breeds of goats have been developed to yield prodigious quantities of product. On a body weight comparison, a good milking goat produces much more milk than a cow in a comparable environment and the Angora goat grows its fine mohair coat at an impressive rate.

On performance alone it is difficult to understand why the goat has not reached a position of importance in agricultural production before now. Even in the developing countries, where over 65 per cent of the world's goats are to be found, they are rarely considered as desirable as cows. However, the situation is changing. Education in developing countries has shown that it is often best to improve indigenous species' breeds and systems rather than import exotic species or breeds from overseas.

In the West the paradox of over-production has forced farmers to seek alternative enterprises concentrating on products for which the demand is still greater than the supply. All of this adds up to an increased interest in the goat and justification for writing this book.

A number of good books have been written by goatkeepers who, because of the lack of professional or scientific advice, are largely self-taught from lessons learned through hard experience. These books are of tremendous help, particularly to the small-scale goatkeeper, but some of the advice is based on subjective observation and experiences with relatively small numbers of goats. In the scientific world, facts must be backed by statistically valid numbers and it is only when this is so that positive recommendations can be made.

This book has been written by drawing on the most recent information published in the scientific literature from around the world and from the experience of working for over 20 years with a large herd of goats at an internationally renowned research institute. No book will ever be complete and this one is no exception. Knowledge about goats is increasing rapidly and therefore no book can be completely up to date. However if this book is used and quoted just a fraction as much as Mackenzie's *Goat Husbandry* it will have fulfilled its purpose, and it is hoped that if there is a demand, future editions will include the latest knowledge about this unique species.

I would like to record my thanks to the many people who, during my long involvement with goats, have freely provided a lot of the information used in this book, and to Dr Angus Russel and Mr Doug Ellis for their constructive criticism of parts of the text. All of those who have provided me with photographs or allowed me to photograph their goats or their facilities are acknowledged in the text.

Special thanks must go to my wife Jacqueline for tolerating the very different way of life since I became a full-time 'goat person'; her hard work has enabled me to spend countless weeks writing. Thanks also to my sons Thomas and Matthew, who have had to own up to their school chums, on many occasions, when asked 'What does your father do?'

ALAN MOWLEM
Reading
March 1988

Foreword

Two goats eye me warily as I turn off the main road towards my home in Suffolk. They are a continuing reminder of the welcome interest in goat farming. It is one of the great advantages of the changes in agriculture that we are seeing a growth in variety, a willingness to experiment, and to move away from the monoculture which has been so characteristic of agriculture since the war.

Many producers have been attracted towards goat farming not only by the demands to diversify but by the increasing interest of the consumer. It started very much as a minority concern: people who found goat's milk more digestible than cow's milk, others with allergic conditions, and some who just relished a change. Increasing foreign travel and the widespread availability of goat's cheese from other countries has now meant that more and more people are willing to try something new. At the same time interest in the specialist production of luxury fibre from goats has been growing. The UK has traditionally been a major processor of cashmere, mohair and now cashgora. There are enthusiasts for the types of goat which can supply this market and already they are showing their wares at a growing number of county shows. It is still very early days to tell whether this is a passing fancy or the basis of a real and continuing business. Nevertheless I hope very much that the pioneers will be rewarded and that we shall be able to make a substantial contribution to the textile trade from home-produced goat fibre.

One of the most encouraging elements in all these developments has been the growing realisation that only the highest standards of hygiene and quality control can ensure that goats' milk and dairy products rise above the rather cranky image which they have enjoyed for too long. That makes the formation and expansion of a number of specialist societies so very welcome. The Goat Veterinary Society, the Goat Producers Association of Great Britain, the British Angora Goat Society and other groups are all contributing to a wholly new attitude towards the goat in Britain. Like so many others, my early views of goats were culled exclusively from childhood tales. They were almost foreign animals, so rarely did we see them in the flesh. My children recognise in Billy Goat Gruff the goats down the road. All of us must welcome this change. The growth of interest has led to a real need for information and advice. Already the Agricultural Development and Advisory Service of the Ministry of Agriculture, Fisheries and Food has responded to this by training specialist advisers. They provide consultancy for farmers and draw on experience gained not only in the United Kingdom but in the countries abroad more versed in modern goat husbandry. This book is another response to the need for information. Alan Mowlem of the Goat Advisory Bureau is fulfilling a real need in this informative and necessary book.

JOHN SELWYN GUMMER

Chapter 1

A Historical Perspective

The goat was one of the first animals to be domesticated by man. Remains have been found in deposits that are 5 million years old. Signs of their domestication have been found in excavations of neolithic sites at Jericho dating from 7,000 BC.[1] As civilisation developed, more evidence of the species appeared and poems and drawings preserved for more than 5,000 years pay tribute to both wild and penned goats.

The constant appearance of the goat in Greek mythology is evidence of its importance at the time of the creation of the many legends. Old Zeus was raised by a goat named Amalthea. He rewarded his nurse by placing her in the sky as the bright star Capella ('Little Goat'). Dionysus, god of wine, was also suckled on goat's milk. Pan, who may have been Dionysus's son, had goat's feet and horns. About 400 BC the Greeks formalised the Zodiac and assigned the position of controller of the Winter Solstice to the goat-fish Capricorn.

There are many references to goats in both the Old and the New Testament; they had become, by Biblical times, important as a source of milk, meat, hair and skins.

At the time of Moses an annual ritual involved the sacrifice of a twin goat as an offering to the Lord while the other was anointed and set free in the wilderness to be a live atonement offering or 'scapegoat'.

By the time of Christ the goat was beginning to be discriminated against and it was seen to be desirable to separate Christ's followers, the good sheep, from the others, the goats. The goat thus gradually disappeared from religious art and writings even though it had played a major part in earlier religious history.

GOATS THROUGHOUT THE WORLD

Today the goat has penetrated to almost every country. The only regions where it is not found in significant numbers are the Arctic countries and the Antarctic. In many countries it is the most important source of animal protein and whole communities depend on their flocks of goats.

[1] See ends of chapters for references.

1

Figure 1.1   A gold and lapis-lazuli figure of a goat nibbling
a tree from Ur, Babylonia about 2,500 BC (courtesy
British Museum).

The total world population of goats is about 470 million.[2] It is impossible
to obtain an accurate figure because goats are not easily counted. They
spend much of their time hidden in the landscape's nooks and crannies
and in steep inaccessible places—unlike cows, sheep and horses which are
usually visible from a long way off. Often those who own goats are
reluctant to admit the size of their herd. About 75 per cent of the goats in
the world are in the developing countries, kept by small family units and
used for the production of meat, skins, milk and hair.

There is still some stigma attached to keeping goats even in those
countries where they are numerous. This can be attributed to misguided
religious prejudice and to the misguided appreciation of Western ideals.
The goat is rarely seen as a status symbol in the way that a cow is and yet

more often than not they make a greater contribution to the welfare of those who own them.

There are many countries and islands with feral (escaped domesticated) goat populations. As far back as the sixteenth century it was appreciated that goats could be useful as a self-renewing source of food for seamen faced with arduous voyages. Many were turned loose on islands through-out the major seaways. These goats flourished and have been added to, over the centuries, by escapees or ones turned loose to eat brush and undergrowth by settler communities.

One of the largest populations of feral goats is to be found in Australasia. It is estimated that there are between 500,000 and 1,000,000 in Australia and about 300,000 in New Zealand. The origin of the latter is credited to Captain Cook who released goats in various locations in New Zealand and the South Seas in the 1770s.

Because of their ability to survive and reproduce in such remote areas many of these feral populations grew rapidly. On Kauai Island in the Hawaiian archipelago 5 goats introduced in 1792 had multiplied to such an extent that in 1850 26,519 goat skins were shipped off the island.[3]

This same ability to survive on remote islands enabled goats to be kept as a source of milk, meat and for companionship by lighthouse-keepers on many inhospitable islands.

These uncontrolled populations of goats contributed to the reputation of the species for deforestation and desertification around the world. The true story, as always, is not simple. In all instances man is the major culprit. The destruction of trees for construction and for fuel has been the biggest influence on deforestation. Often goats were around to browse through what was left. Also badly managed livestock, including goats, were allowed to destroy forests or the forest was cut down to provide grazing. In such situations the goats were always the last to remain thanks to their unique ability to be able to find sustenance where other species often could not.

Undoubtedly large populations of unmanaged goats, in some fragile environments, particularly islands, have done considerable damage but this is equally true of many other species turned loose by man.

Over the last few decades the importance of the goat in agricultural production has gradually emerged. Studies have been carried out to establish the true role and effect of the different ruminant species. It has been found that goats and other ruminant species can complement one another when grazed together. Land grazed by cattle and goats produced a 25 per cent greater return than when the same land was grazed by cattle alone. This is largely due to the goats' selective grazing and browsing of plant species which are not eaten by cattle and which often inhibit grass growth or reduce the area accessible to the cattle.

Figure 1.2 Goats coming in to be milked in Sweden (courtesy of B. Forsberg).

Goat farming is most developed in France where more than one million goats are kept for milk which is used almost exclusively for the production of cheese. Many of the farms are family units with typically up to 200 or 300 goats. In some cases the cheese will be made on the farm and in others milk will be supplied to large co-operative dairies. Many of the cheeses are particular to certain regions and have become well known and appreciated in a similar way to some of the regional wines.

GOATS IN BRITAIN

There is a long history of goat domestication in Britain. Although remains of goats have been found in levels from the Pliocene period the indigenous domesticated British goat, now also disappeared, almost certainly originated from animals brought to these Islands by settlers in pre-Roman times. The breeds we have today are the product of comparatively recent crossings of these indigenous goats and a number of imported breeds.

Through the ages the goat in Britain has traditionally been a creature of poor regard, persecuted rather than praised, with the less desirable side of its character being emphasised far more than the attributes that make it useful. Much of the blame for this can probably be levelled at the church for the way it has treated the goat and the way the people were gradually indoctrinated into believing that the goat was undesirable or even evil.

The goat was used as a pagan symbol and by the Middle Ages, during the period of witch hunting and persecution, it was associated with witches. Gradually all of this prejudice culminated in the nineteenth century with the goat and the Devil being classed as one. The popular image of the Devil is still of a shaggy creature with goat's horns, a beard and cloven hooves.

The independent goat did not take to the enclosure revolution that so changed agriculture in the Middle Ages. It was and still is a difficult animal to contain and this, coupled with its destructiveness when not well managed and with all the theological disapproval already mentioned, virtually removed it from the mainstream of British agriculture. It was only those whose independent character matched that of the goat who kept on with their goat-keeping, usually in situations and circumstances where it would have been impossible to have supported sheep or cattle.

Through such diehards the British goat continued. Some escaped and some were abandoned when remote communities were forced to leave their islands or villages. These goats thrived and we still have surprisingly large feral herds in many of the more remote regions of the British Isles.[4]

By the latter part of the last century the interest in animal breeding and improvement had embraced even goats and they were being selectively bred for both looks and for production. A lot of interest stemmed from the wives and daughters of the men pushing forward the frontiers of the British Empire. Of these ladies many were left at home with often a good education but in a society that frowned upon working women. Those with country backgrounds or leanings found the goat an ideal subject with which to develop this interest into small-scale farming and food production. Thus the pastime of goatkeeping and goat breeding became acceptable and even today is largely the province of women.

One of the first steps to improve the quality and production of the indigenous goat was the result of the regular traffic between England and the growing Empire. Many of the ships travelling back from the East carried goats to provide fresh milk for the returning colonialists who had become aware of its value. These goats on arrival ended up in zoos, animal collections or at the houses of those retiring from government service overseas. Much interest in crossbreeding resulted, using this mixture of breeds along with some from Switzerland and in 1879 the British Goat Society was formed.

This evolution of the interest in goatkeeping has meant the goat has always, until recently, been regarded as a creature of the back garden or pony paddock and it has been bred to be productive and with good conformation in very specialised surroundings bearing little resemblance to a truly agricultural environment.

Nevertheless the enormous production performance of some of these goats does indicate the potential of the species. It follows that goats capable of producing 1,500–2,000 kg of milk per year in very specialised environments ought to produce progeny that would be capable of producing at least 1,000 kg in a more typical farm environment, a yield that larger-scale goat farmers would be very pleased to average.

During both world wars the interest in goats was accelerated as part of the campaign to encourage home food production. However, particularly after World War II, the goat was largely ignored as the nation rebuilt its industrial and agricultural production. During this period farming underwent a massive transformation as new technology and mechanisation favoured larger farm units. The goat, still suffering from the false reputation of previous centuries, did not really fit into this scene at all. With the exception of one or two individuals, commercial goat farms were unknown and the average herd size of the thousands of British Goat Society members remained at two or three animals.

This situation prevailed until the 1970s when a number of factors were causing a certain amount of reappraisal of food production from British agriculture. Goat milk had always been recognised by many people as a useful substitute for cow milk when this seemed to cause allergic responses in infants. This demand resulted in a small haphazard trade in milk, mainly from small producers who sold it directly from the farm or more typically from the back garden.

During the 1970s the traditionally conservative British were beginning to travel more and more and the increasing political and economic links with the rest of Europe encouraged many to holiday on the Continent. These adventurous holiday-makers soon developed a taste for Continental foods, and what had, until then, been a street-corner delicatessen trade spread into the large retail supermarkets. Among these 'Continental' products was a variety of goat's cheeses and yoghourts.

All of this went hand-in-hand with an increasing awareness of how badly we were feeding ourselves, as a nation, and how badly our health suffered as a result. The healthy image of some of the fresh yoghourts, lactic cheese and the low-fat dairy products attracted even more attention thus accelerating the boom in sales. All this added up to a rapidly developing market for professionally produced and presented goat products.

This development of the market for goat products coincided with the over-production of many traditional products from European agriculture. The resulting quotas on cow-milk production and the falling prices of many other commodities have resulted in many farmers anxiously looking for alternative enterprises producing products for markets that are not yet over supplied. In the last few years many commercial-scale goat units have appeared and more recently the production of fibre from goats (mohair and cashmere) has created even more interest in the species.

All of this activity has led to the formation of many new organisations. The British Goat Society has flourished but has been rather tardy in keeping up with the pace of the interest in large-scale goat farming. As a result the Goat Producers' Association was formed in 1984 to foster the interest of those developing commercial goat enterprises. Also in recent years the Goat Veterinary Society was formed, a very welcome step and evidence that the veterinary profession recognised the increasing importance of the goat.

The company Caprine Ovine Breeding Services Ltd (COBS) was formed in 1981 to work towards the development of a goat artificial insemination service in Britain. This technique has tremendous potential value and could eventually revolutionise goat breeding in the same way it has for cattle.

The current interest in Angora goats has brought about an increase in membership of the British Angora Goat Society from 50 to over 1,400 in less than 4 years. In 1986 the Scottish Cashmere Producers' Association was formed to further the interest in production of this other valuable goat fibre.

It is a strange, but not unfamiliar, paradox that all this has been going on at a time when the British Government has been cutting the funding of agricultural research and development. These cuts have resulted in almost total curtailment of what little research work was being carried out on goat production. This short-sighted policy will obviously slow up the potential development of a British goat industry which is making a significant and increasing contribution to agricultural production. The demand for goat products in Britain is, at present, greater than the supply, and in addition there is the possibility of an export demand for some products.

REFERENCES

1. Zeuner, F. E. (1963), *A history of domesticated animals* (Hutchinson, London).
2. Food and Agriculture Organisation (1984), *Products Yearbook*, Vol. 38 (Food and Agriculture Organisation, Rome).
3. Dunbar, R. (1984), 'Scapegoat of a thousand deserts', *New Scientist*, 15 Nov. 1984.
4. Whitehead, G. K. (1972), *The Wild Goats of Great Britain and Ireland* (David & Charles, Newton Abbot).

Chapter 2

Goat Characteristics and Breeds

The goat is, taxonomically speaking, in the order *Artiodactyla*, the family of *Bovidae* and the genus *Capra*. Also in the genus are the ibex, the bezoar, the aegagrus and the markhor. The markhor and its subspecies have been extensively hunted and all are now on the International Union for the Conservation of Nature (IUCN) list of threatened species.

Scholars attribute five or six wild species to the genus *Capra* (the goats) and there is still some discussion over which species is the origin of the domesticated goat. It is conceivable that more than one could be although it is now generally agreed that the bezoar *Capra hircus* and the subspecies *Capra hircus aegagrus* are the most likely candidates. These species are to be found in Turkey, Iran, Southern Turkmenistan, Western Afghanistan and some Greek Islands.

The ancestors of the European domesticated breeds were thought to have come into Eastern Europe already domesticated from South West Asia. Remains have been found in Neolithic Swiss lake-dwellings and from sites in North Eastern Yugoslavia and Hungary.

The domesticated goat, *Capra hircus*, is now found throughout the world in many forms—there are in fact well over 200 identifiable breeds of goats. The question of the difference between sheep and goats is often raised and indeed with some breeds of goats it is difficult to differentiate them from some breeds of sheep. The most telling difference is not visible, as sheep have 54 chromosomes and goats have 60. Visible anatomical differences may not be too convincing between some breeds. Goats generally hold their tails up whilst the tails of sheep hang down. All male and some female goats have beards, a feature only seen unconvincingly in some primitive breeds of sheep. Male goats also have a characteristic smell which is quite different from the smell of a ram and rams have a secretory gland on the hind feet which goats do not possess.

It is in behaviour that goats differ most from sheep and for that matter any other livestock and in various early textbooks on animal husbandry and breeding there are some splendid accounts of the goat's behaviour which still 'ring true' today. F. Finn, FZS in Hutchinson's fortnightly magazine *Living Animals of the World* published in the early 1920s states: 'The difference between the temperament of sheep and goats is very curious and persistent, showing itself in a marked way, which affects their use in domestication to such a degree that the keeping of one or the other often marks the owners as possessors of different degrees of civilisation. Goats are restless, curious, adventurous and so active they cannot be kept in enclosed fields. For this reason they are not bred in any numbers in lands where agriculture is practised on modern principles; they are enterprising and destructive.'

The American author Robert Wernik wrote recently 'Sheep are conformists, goats are unpredictable, flighty and capricious. . . . If a sheep hears a low flying plane it will become frightened and likely to run whereas the goat will stand and watch. . . . Goats are basically down-to-earth creatures with a genius for making the best out of any situation they may find themselves in. And centuries of cohabitation with mankind have put them in all kinds of situations; they have learned how to survive them all.'

Perhaps the most erudite description of all is by Professor Low in his *Domesticated Animals of the British Island* published in 1845, a standard work

Figure 2.1 A bezoar at London Zoo.

for rare breed historians. He says 'the goat although obeying the law to which all the domesticated animals are subject, and presenting itself under a great variety of aspects, retains many of the characters and habits that distinguish it in the state of liberty. It is lively, ardent, robust, capable of enduring the most intense cold and seemingly little incommoded by the extremes of heat. It is wild, irregular and erratic in its movements. It is bold in its own defence, putting itself in attitude of defiance when provoked by animals, however larger than itself. . . . When the Goat is kept apart from the flock he becomes attached to his protectors, familiar and inquisitive, finding his way into every place and examining whatever is new to him. He is eminently sociable, attaching himself to other animals however different from himself.'

Much more recently Jacky Gillott in her book *Providence Place* gives her version of the goat's character. 'There is something ancient and inscrutable about the goat. Its yellow eye with his black, oblong pupil gazes at you out of forgotten time. It possesses a simple minded, some would say wilful, streak which makes you sense that, like a cat, it has a clear memory of its own wild origin and that your proprietorship is something it is prepared to tolerate, to trade for a little extra food perhaps, but only on its own terms. Goats respond well, affectionately even, to those they judge worthy of them. Towards others they demonstrate lighthearted disregard which is, I'm sure, the source of the furious dislike that so many who have had contact with them seem to feel for goats. For such people the goat's yellow stare is a terrible and baleful beam. It is the confident sneer of one's superior.'

Although in their behaviour goats seem to be very different from sheep or cattle, and because of this they are treated with a mixture of disdain and horror by some cowmen and shepherds, they do have many similarities with those two species. They are all ungulates or cloven-hooved, they are all ruminants, they have similar dentition, there are dairy breeds of all three species, there are horned and hornless breeds and all have been domesticated for thousands of years.

Of the vast number of identifiable breeds of goats there are eleven in Britain with several other crossbred types that are not yet recognised as breeds. Quite rightly it is difficult to import new breeds into this country because of the strict control over animal health imposed by the Ministry of Agriculture. Being an island we have a good opportunity of avoiding many of the diseases that are endemic in many of the large continental land masses.

Thus a good health status with long established pedigrees creates a demand for UK goats overseas and has resulted in a thriving export trade in live goats and for the first time, in 1988, in frozen semen. It is now possible to find British Saanens, British Toggenburgs, British Alpines and

Figure 2.2   Newly arrived British Saanen kids on the island of São Tomé, West Africa (courtesy of Agricultural Export Services).

Anglo-Nubians in such countries as China, Mauritius, India, East Africa, the West Indies and requests for goats are even coming from other European countries such as France and Holland.

Of the eleven breeds found in Britain none can be called native. Most are the result of importations during the last 150 years and a good deal of cross-breeding with these imported breeds.

SAANEN

The Saanen is a Swiss breed originating from the Saanen Valley in the south of the Canton Berne. In the Simmental Valley, where it is also found, it is known as the Gessenay goat.

It is a white goat with sometimes black pigmentation on the skin which is most visible as spots or blotches on the udder and ears. Mature females will weigh about 60 kg and mature males about 85 kg. Females often have beards and tassels (or wattles) although neither are essential features of the breed. Horns may be present although there has been rigorous selection for the polled condition in their native Switzerland. The Saanen is one of the best milk producers with yields approaching 2,000 kg not being uncommon.

The first Saanens to be introduced into Britain were those imported by a Mr Hook at the end of the last century and by a Mr Hughes a few years later. These early imports were not kept pure and the pure Saanens found in Britain today are nearly all descendants of some imported from Holland by Mr Palmer and Mr Hughes on behalf of the British Goat Society in 1922. Few other importations were made, the last being in 1967.

THE BRITISH SAANEN

As its name suggests the British Saanen is derived from the Saanen and in particular those imported at the early part of the century. These were crossed with other breeds and from these came an improved Saanen type which was named the British Saanen, a breed recognised by the British Goat Society since 1925.

In appearance the British Saanen is very similar to the Saanen and it is often difficult to tell them apart. However the British Saanen is generally larger with females averaging 70 kg and males about 90–100 kg.

In terms of milk yield it is unsurpassed having outstripped its parent breed and all others. The world record milk yield for a goat is from a Saanen in Australia of British lineage. This goat, Osory Snow Goose,[1] produced 3,300 kg of milk in her first lactation and 3,506 kg in her second. Although such enormous yields can only be produced by a great deal of individual attention with respect to feeding and are unlikely to ever be achieved in a farm situation, it does show what tremendous potential the breed has for high yields. It is obvious that British Saanens, in a farm environment, ought to be able to average 1,000 kg per lactation which is the target set by most commercial dairy goat farmers.

Another trait of the Saanen type that makes them particularly suitable for intensive farming is their docility and placidness. In a contented closed herd, where all goats are related or used to each other, serious fighting, bullying, jumping out of pens and many of the other bad habits attributed to goats will be unknown.

THE TOGGENBURG

In terms of performance the Toggenburg is very close to the Saanen and like the Saanen it has given rise to an improved British derivative. The Toggenburg comes from the Canton of St Gallen in Switzerland and has been exported all round the world.

It is light brown in colour with characteristic white stripes on the face, white lower legs and white around the tail. Sometimes its coat is quite long with some animals having a 'skirt' of long hair almost covering the upper parts of the legs. Also like the Saanen the males and sometimes the females have beards and horns although attempts have been made to breed these out.

It is a smaller goat than the Saanen with females averaging about 60 kg and males about 75 kg. While not achieving quite the high milk yield of the Saanen they are good milkers and do produce better quality milk in terms of butter fat and total solids.

THE BRITISH TOGGENBURG

The origins of the British Toggenburg are similar to those of the British Saanen. It is the product of cross-breeding during the latter part of the last century and the early part of this one when pure Toggenburgs were

Figure 2.3   British Toggenburgs at Nut Knowle Farm.

crossed with various types until a fixed type was produced which gained recognition with the BGS as the British Toggenburg in 1925.

The British Toggenburg is about 10 kg heavier than the Toggenburg with a heavier milk yield. Milk quality is also better than in the pure breed. It retains the same colouration and markings but tends to have a shorter coat. It has a reputation of being an excitable breed with a propensity for getting out of most enclosures. These unfortunate characteristics do not usually make it first choice for large-scale commercial units but exceptions are to be found and those who do keep them in large numbers usually say that the improvement in milk quality over the British Saanen outweighs the few problems caused by the difference in temperament. A number of commercial herds mix British Toggenburgs with Saanens to improve the quality of the overall milk production.

THE BRITISH ALPINE

When discussing the British Alpine it is important to emphasise the British prefix because there are Alpine breeds in other European countries that look nothing like it. Although the British Alpine is marked almost exactly like the Swiss Bundner Strahlenziege breed it has only a very distant relationship with it, being a created breed with origins going back to a single Swiss female goat imported with the 1908 goats. This produced kids that were used to create the fixed colour pattern and conformation we see today.

The British Alpine is a handsome goat with a coat marked in a similar way to the Toggenburg but with short glossy black hair. It is about the same size as a British Toggenburg. It has a reputation for being a little hardier than the other breeds of Swiss origin and may be suited to the less intensive enterprises. It does not average such good milk yields as the British Saanen or British Toggenburg but the milk quality is good.

THE ANGLO-NUBIAN

As its looks suggest the Anglo-Nubian is not of Swiss but of Eastern origin. It is the product of crossing the indigenous English goat with the Zaraibi from Egypt and the Jamnapari from India. There is also a little Swiss blood in the breed as well. It is a big goat with females usually averaging over 70 kg and males can be more than 100 kg. Their distinctive Roman nose and large pendulous ears are testimony to their origins. They can be any colour and the British Goat Society has a most interesting list of colour descriptions for the purpose of registration.

Figure 2.4   British Alpines (courtesy of the Food Research Institute, Heading).

They generally produce less milk than the Swiss breeds but their milk is of significantly better quality in terms of both fat and protein. This means they are favoured by those who produce cheese as the greater yield of cheese more than compensates for the lower milk yield. Apart from milk production their big body frame and relatively muscular conformation suggest they may be one of the best breeds we have for developing meat production.

Anglo-Nubians are an excitable breed and always give the impression of being neurotic and highly strung and probably not suited to large intensive herds. However, those who enthusiastically keep large numbers say they are no more of a problem than the Swiss breeds.

Another leftover from their semi-tropical heritage is their longer breeding season. Goats in and near the tropics are not normally seasonal breeders and Anglo-Nubians have not entirely lost this characteristic. They usually come into season at least a month before the other breeds and may continue to cycle for a month or two after. Because of their

Figure 2.5   Anglo-Nubians.

Figure 2.6   Angora goats at Pound Farm, Devon.

origins they are suited to semi-tropical environments and there is quite a good export demand for them.

THE ANGORA

Until comparatively recently all goats in Britain were dairy breeds or derivatives from them. In 1981 a small number of Angoras were imported, some from New Zealand by Mrs M. Rosenberg and Mr J. Norris and some from Tasmania by Miss P. Evers-Swindell. From this small number has grown a vigorous interest in the breed and in mohair production as an alternative farm enterprise. Angoras originate from Turkey and it was not until the mid-1880s that any were allowed out of that country. They are now to be found in large numbers in South Africa, the USA, Argentina, Australia and New Zealand.

The Angora is smaller than most dairy breeds with mature females in Britain averaging about 45–50 kg with males around 60–65 kg. Their most striking characteristic is their long coat of fine lustrous mohair. This grows at the rate of about 2.5 cm a month and therefore their appearance depends on when they were last shorn. Angora goats are usually white and normally these would be selected against coloured goats. However, brown and black Angoras do exist. They normally have large horns with the male's sometimes reaching a spread of over 1 m. In general shape Angoras look more like sheep than goats and it has been found that if crossed with dairy breeds they can produce kids with good carcass conformation.

THE GOLDEN GUERNSEY

As its name suggests this goat comes from the Channel Islands although it almost certainly has origins in Mediterranean countries as there are remarkably similar goats on Malta and in Spain.[2] In appearance they look like ginger or sometimes reddish brown Saanens, although their coat is generally longer with some having a long skirt of hair around the legs like some pure Toggenburgs.

They have not yet been developed or improved like the other dairy breeds and because of their small numbers improvement is slow. They are not particularly good milkers although a few animals are showing that they may have potential in this respect. They seem to have a placid nature similar to Saanens.

THE BAGOT

The Bagot is most definitely a rare breed. It is almost certainly related to the Schwarzhal breed of Switzerland even though the British herd had remained closed since its establishment in the fourteenth century in the park at Blythfield Hall in Staffordshire, the home of the Bagot family.

They are fairly small goats, compared with the dairy breeds, with a long shaggy coat which is black over the head and shoulders with the rest pure white. They have horns, those of older males often reaching a length of almost 1 m.

The herd had become very inbred and their survival was in doubt until the Rare Breeds Survival Trust organised a breeding programme where some were crossed to similar breeds to improve their general health and condition. Some have now been relocated on rare breed farms. As yet no commercial value has been identified; in spite of their long hair they do

Figure 2.7   A Bagot male.

not seem to yield much cashmere. As with all rare breeds there is a very good case for preserving them because they may have traits that may, one day, be considered useful. This is apart from the aesthetic pleasure of seeing unusual minority breeds.

THE WEST AFRICAN DWARF GOAT

This breed, which includes the Nigerian Pygmy breed, has no commercial value in this country although these are good milkers for their size. They can be almost any colour and usually have horns. They are very small with mature females weighing about 30 kg and males about 35 kg. They are very popular on farm parks and in children's zoos.

CASHMERE GOATS

There is no such thing as a cashmere breed although in various parts of the world goats have been selectively bred for cashmere production. Cashmere is the fine hair grown as insulation by some goats in cold environments. Goats producing the most cashmere are to be found in China and other countries bordering the Himalayas. In recent years feral goats in Australia and New Zealand have been used as the base breed for developing cashmere producing goats. Some of these have been imported into the United Kingdom.

Work is also being carried out at the Macauley Land Use Research Institute using ferals as the base breed, to develop a cashmere- and meat-producing goat to graze on unimproved uplands.[3]

FERAL GOATS

There are surprisingly large populations of feral goats in the more remote parts of Britain such as the Highlands of Scotland, North Wales and the North of England.[4] They are the product of escapees from remote communities and in most cases their history goes back at least 100 years. They are usually small, hairy and very hardy but with poor production. When they freely interbreed a greyish brown colour is fairly typical but where groups have remained more or less isolated they may be more distinctively marked depending on their origins. One such group is the white goats on Great Orme in North Wales. These are said to originate from the Windsor Whites which were cashmere goats given to Queen Victoria and used for Welsh regimental mascots. Some of these can also be seen in Whipsnade Zoo.

Figure 2.8   Feral goats from the Cheviot Hills at Ash Farm, Devon.

ACQUIRING GOATS FOR THE FARM

For those wishing to farm goats on a commercial scale the first choice is, of course, whether the enterprise is going to produce milk or fibre. If it is to be a milking unit it can now be seen that there is some choice of breed, each having its good and bad points. After considering production parameters such as milk yield and quality and management factors such as temperament and ability to jump, the final decision will probably also be influenced by personal preferences. Many people just like certain breeds for often inexplicable reasons.

The choice having been made, the goats now have to be found. It is possible to acquire goats from the main dairy breeds, in ones and twos, for comparatively low prices. However, it should be realised that mature goats usually come onto the market for very good reasons such as poor breeding, milking performance or because of some vice that cannot be cured. If a herd is built up from this type of goat it would be necessary to operate a rigorous culling policy during the first few years to remove the poor producers and those that have difficulty in fitting into the herd. Thus what may have been an inexpensive nucleus now has to carry the cost of the animals that have been culled plus the cost of keeping them until they were culled.

Goats are very hierarchical animals and take a long time to settle down if unfamiliar ones are put together in a typical yard or pen system as favoured for large-scale milk production. Some, probably those which have been pets, may never settle into the way of life of a big herd. During this settling in and culling out period, yields will be low and animals capable of yielding 1,000 kg or more of milk may drop to a third of that or may even dry up altogether.

If these problems are considered to outweigh the advantage of starting with a ready-made herd of adult milking goats the alternative is to start with young goats. This suits some people because it gives them a chance to get to know the goats' requirements and their differences from other farm stock before the problems of milking have to be faced. Often the young goats can be acquired during the period when the milking parlour is being set up.

In this situation it would be normal to buy young females of 4–6 months of age in the early autumn. If they grow well enough they can be mated later that year and will be kidding and producing milk during the late spring or early summer of the next year. Goats under one year of age generally settle down together without a lot of problems.

It is a rather different situation with fibre-producing goats. These are kept much more extensively in an outdoor environment where space is not a problem. The only time struggles for dominance will be a problem here will be when there is competition for food trough space or for shelter space. Angoras have a reputation, among those new to them, of being aggressive, bossy goats. This is almost certainly due to the fact that almost all Angora herds in Britain, at the moment, are relatively new with goats being added all of the time. Almost certainly when herds become established with few new goats being added, this behaviour will change.

When purchasing fibre-producing goats the choice will almost certainly be one of simple economics. If inexperienced in judging the quality of the goats it would be wise to enlist help. Even if this involves a fee it could result in the avoidance of expensive mistakes. Once the quality of the goats is established it is then a question of value. Will it be possible to achieve a reasonable return in terms of the fibre and kids that it will produce? If the sale of stock is to be a significant part of the enterprise then the pedigree and quality of female stock may be more critical than with purely commercial fibre production.

For commercial fibre production, a flock made up of females of average quality, may be improved considerably if first-class males are used.

With both cashmere and Angora goats it should be possible for a vendor to produce evidence of the yield and quality of the fibre produced at the last clip for any animal. With Angoras much emphasis is put on the result of fleece tests. These are a useful guide but no more. The quality of a fleece sample will vary according to the site it was taken from and the nutritional and health status of the goats. There have even been cases of kemp being picked out of a mohair sample before the sample was analysed! It is unwise to choose male Angoras for stud until they are 18 months old as their fleece will not mature until that age. It is, therefore, not possible to judge his fleece quality until then. New stud males have even been bought with only one testicle, an obvious case of an inadequate presale inspection.

It is vital that any goat is properly examined before purchase. The prospective purchaser must get in the pen, must feel the goat, must look for anatomical defects and in the case of fibre goats must feel the fleece and part it to closely examine it for defects. It is also important, particularly with milk breeds, to see the goats walking around so that one can see if there are any problems such as narrow or weak back legs that would be undesirable in a potentially good milker.

REFERENCES

1. Jameson, G. and C. J. (1978), 'Osory Snow Goose', *British Goat Society Year Book 1978* (British Goat Society, Bovey Tracey).
2. Menos, C. E., and Tejon D. T. (1980), *Catalogue de razas autoctones Espanoles 1. Espeies ovina y caprina* (Ministeria de Agricultura, Madrid).
3. Russel, A. J. F. (1987), 'The beginnings of Scottish cashmere', in *Scottish Cashmere*, eds Russel A. J. F., and Maxwell, T. J. (Scottish Cashmere Producers' Association, Edinburgh).
4. Whitehead, G. K. (1972), *The Wild Goats of Great Britain and Ireland* (David and Charles, Newton Abbot).

Chapter 3

Housing and Fencing

It is important, if there is a choice, to site buildings for housing milking goats near to the parlour or at least where an uncomplicated journey to and from the parlour will be possible.

Much frustrating effort, retrieving goats from areas or buildings into which they should not have access, can be avoided if the layout of buildings is carefully thought out. Storerooms are the biggest problem as goats gaining access to these may find feedstuffs or harmful chemicals which at the very least may prove expensive and at the worst may prove fatal.

THE BUILDING

Pen Construction

When designing or constructing pens for goats it must be borne in mind that they are more active, agile and inquisitive than any other farm species. This can mean destruction of pens and will certainly mean escape if the pens are of inadequate design or construction.

The most goat-proof pens are of brick or block construction. Timber may be damaged by chewing and metal hurdles or similar barriers may be climbed through or over or jumped. Goats may use features such as window sills as 'launching-pads' for jumping out of pens.

The height of the pen sides will depend to some extent on the breed and character of the goats. Saanens are generally less likely to jump out of pens, Toggenburgs and Anglo-Nubians often will. If the goats are well fed and content they will be less likely to jump and young goats are more likely to jump than old ones.

A pen height of 1.5 m will keep in all goats providing there are no stepping-stones or launching-pads. Contented goats with little inclination to jump can be contained behind barriers of 0.91–1.1 m. Standard tubular sheep-penning systems can be used in this instance but it must be remembered that goats raise themselves up on their hind legs and rest their front legs on the tops of pen sides. This can result in feet or legs becoming

23

Figure 3.1   Timber damaged by chewing.

trapped between pen divisions. This can be avoided with permanent pens by adding a horizontal bar to fill the gap.

Door and gate latches must be carefully thought out as goats are able to undo many conventional latches and bolts. Figure 3.2 shows a modification to a drop latch that will prevent goats opening it.

If deep-litter bedding systems are used pen barriers should be adjustable so that they can be raised as the depth of bedding increases. Bedding collected over a 3–4 month period may be 0.5 m deep.

Male Housing

The most important point to remember when considering buildings for males is that during the breeding season they smell very strongly indeed. This means the male building must be sited away from areas where the smell could be a problem, e.g. the milk-handling area or domestic dwellings.

Figure 3.2   A goat-proof gate latch.

The environment required for males is more or less the same as for females except some breeds do seem to suffer with arthritic conditions and, therefore, care should be taken to make sure the males are in a dry atmosphere. Some farm buildings can have condensation problems.

Males are, of course, much stronger than females and their pens must be more substantial. Males have been known to break through gates made from 1.5 in diameter galvanised pipe.

Kid Housing

If kids are artificially reared consideration must be given to their housing. As long as they are not too cold many farm buildings will be suitable. It is preferable for kids newly separated from their mothers to be housed out of hearing range, if at all possible, as this will reduce stress and both groups will settle down in a few days. In some environments this may not be a problem at all. The important criteria for kid housing are:

1. Freedom from draughts.
2. Airy with good ventilation.
3. Not too cold (optimum 12–18°C).
4. A clean and hygienic environment.

It is possible to satisfy these points in most enclosed farm buildings. To be adequately protected from draughts kids less than 8 weeks of age should be housed in pens with at least two solid sides and not facing towards any opening exposed to winds from outside.

As kids are usually masters at getting out from pens it is likely that solid

Figure 3.3   A heater lamp over a kid rearing pen.

sides all round will be preferable during the period when they are small and more susceptible to adverse environments.

If cold is the only problem in an otherwise suitable building localised heat can be provided by suspending a board, galvanised sheet or similar material over part of the pen. An infra-red lamp or similar heat source can also be suspended to create a warm spot where kids can lie (Figure 3.3).

It is important that any heat source, such as an infra-red bulb, is protected by a wire guard because active kids may well break bulbs by jumping up and resting their front feet on them. Once the kids become very active in this way it can be considered that they are old enough to do without the additional heat.

Once they are weaned kids can be moved to a more open environment as recommended for the adults.

BEDDING

The natural cleanliness of goats can be preserved when they are housed indoors by providing them with some clean bedding material. In some systems where the goats are permanently housed this bedding will be continuously added to over long periods. This deep-litter system has a number of advantages. The increasing depth of bedding is very absorbent, the rotting layer at the base generates heat and of course by only cleaning out the bedding a few times a year the system saves labour.

A variety of materials can be used as bedding. Wheat straw is useful as it is clean, dust free and is not eaten too much by the goats. Other straws are acceptable but barley in particular contains a lot of dusty, spikey bits.

Almost any dried plant material could be used for bedding. Goats given the opportunity to sort through large racks of poor quality forage such as bean straw (see Chapter 4) will pull a lot out onto the floor and this will act as bedding. Inedible, musty or spoilt hay can also be used as bedding but the seeds are a disadvantage if the bedding is subsequently spread on the land.

A self-draining floor of rammed hard-core is cheap and tends to require less bedding than concrete. Concrete does have the advantage of being washable if this is considered a necessary part of the hygiene routine.

FEEDERS AND FEEDING SYSTEMS

Goats in loose pens will need provision for feeding forage and possibly a concentrate feed. A range of feeders or racks for conserved forages such as hay or straw can be bought or indeed made.

Goats are generally wasteful feeders of forage because they spend much time picking through the material selecting out particular bits. In doing this a lot of material will be pulled out onto the floor and once soiled will not be eaten.

Various forage feeders have been designed to overcome this problem. Two basic design types have evolved. One is where the goat puts its head through a hole which is usually shaped so that it cannot pull its head back unless it raises its head. This principle can also be used for central feed barriers either side of a feed passage (Figure 3.4). The other type is more conventional and consists of a mesh or slatted rack that is fixed to a solid wall or pen division. If the mesh or slats are narrow enough (38 mm/ 1.5 in) wastage will be partly controlled. An improvement on racks where forage is pulled out can be made by fixing a tray to prevent this material becoming soiled.

There is an additional problem with goats kept for their fibre because pulled-out forage, seeds, etc. can contaminate the fibre to such an extent that its value will be considerably downgraded.

Feeders for poor quality forages such as straw and pea and bean haulm

Figure 3.4   Goats either side of a central feed barrier at Nut Knowle Farm, Sussex.

need to be very large if the goats are to be given the opportunity to sort through the material picking out the more digestible parts (see Chapter 4).

Whichever design of forage feeder is used and whichever breed of goat is kept there will always be at least one animal that will get into the feeder if it is at all possible!

Hay nets are not a good idea for goats, particularly kids, as they will jump up onto them and legs may become entangled and sometimes broken.

Concentrate feeders

The simplest concentrate feeder is a bucket and the next simplest is a trough. In other words almost any container can be used but as with forage feeders some designs are more efficient than others.

Once again waste can be a problem, especially when coarse mixes are offered. The ideal feeder is one that cannot be fouled by faeces or from mud and dirt from feet. This can be achieved by using the same principle as for forage, i.e. by using a barrier through which the goats put their heads when feeding.

This can be in the form of vertical or diagonal bars or specially shaped 'tombstone' barriers (Figure 3.5). This system reduces spoilage to a minimum and also has the advantage of it being unnecessary for anyone to go

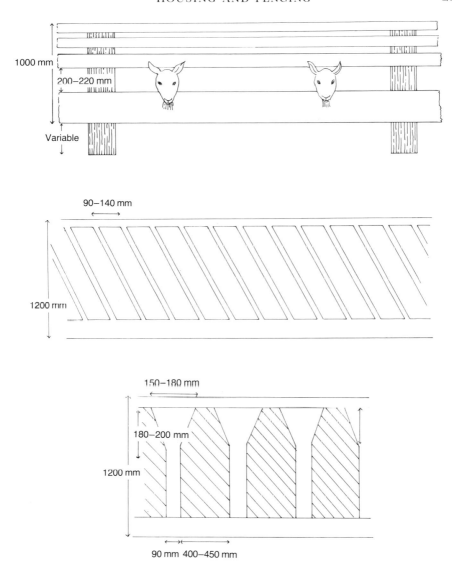

1000 mm

200–220 mm

Variable

90–140 mm

1200 mm

150–180 mm

180–200 mm

1200 mm

90 mm 400–450 mm

Figure 3.5 Examples of feed barriers.

into the pen to dispense the feed. Anyone who has done this will know that one becomes overwhelmed by the goats and there is then a tendency to dispense the feed all over the floor! Feed passages for both forage and concentrate should be 2.5–4.5 m wide depending on the method of dispensing the feed. The narrow passage would be used in conjunction with a tractor and fore-end loader and the wider passage would be for dispensing from a forage box.

Figure 3.6 illustrates another solution to this problem. Here a trough is used which is fixed into a gap in the outside wall of the goat shed in such a way that it can be swung on a pivot so that it can be filled from outside the pen and then tipped back to allow the goats to feed.

A system used in some of the larger intensive units in France involves a conveyor-belt-type feeder which moves down a central passage between two pens. The goats put their heads through a barrier and feed from the forage or concentrate as soon as it reaches them.

If goats are kept in single pens buckets for food and water are usually attached to the outside of the pen gates and the goats put their heads through a hole in the gate to feed. This system is not suitable for horned goats. A small hay rack is usually fixed to the side of the pen.

WATER

Goats need clean water at all times, with a high yielding milker requiring up to 18 l/d. Most self-filling troughs and bowls are suitable for goats although individual animals may have peculiar preferences and may, therefore, refuse to drink from a particular vessel.

To avoid water becoming contaminated with faeces, etc. it is a good practice to raise the trough up to a height of about 1.2 m and to provide a step up with concrete blocks. This allows access for drinking but there is then less tendency for faeces to be dropped into the water.

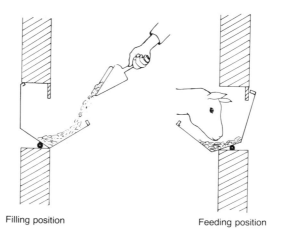

Filling position   Feeding position

Figure 3.6   A hinged trough that can be tipped outside the pen for filling.

OUTSIDE FACILITIES

Many large milking herds are yarded or zero-grazed and, therefore, will need pens similar to those described with, perhaps, access to a separate yard from time to time for additional feeding or for servicing the main holding pen.

Many goats, however, will be turned out to graze and in fact for cashmere and mohair production this is necessary to be economic. Dairy goats, if grazed at all, would probably be turned out for about 7 months of the year. Fibre-producing goats would be left out for as long as the condition of the ground allowed. In the north snowfall would be another factor.

There is much debate and disagreement about goats' need for shelter. In general the improved dairy breeds are not very weatherproof. They do not have the same insulating fat layer as sheep and they do not have a thick hide like a cow. The main consequence of this seems to be a real dislike of cold and wet together and if pushed to extremes in these conditions it is likely that the goats would become chilled.

Angora goats do have a long coat of mohair fibre and they do tend to have more subcutaneous fat than dairy breeds. However, as they are a

Figure 3.7 Angoras and Angora cross-breds winter housed in a polypen.

relatively new breed to this country we still do not yet have enough experience to know just how well they can tolerate the extremes of the British climate. Some Angoras—particularly, but not only, those from Texas—do have a greasy fleece and it seems likely that such a fleece would be more waterproof than a soft dry one.

Because Angoras originate from and are kept in large numbers in dry climates it has been assumed that they will not do well in wet climates. All the evidence so far suggests that, like other goats, Angoras are very adaptable. It is now believed that in all but the coldest, wettest environments Angoras need no more than simple field shelters if natural shelter is not available.

Many cashmere goats are derived from feral animals and will still have much of the hardiness of this type of goat. Feral goats exist and indeed thrive in some of the remotest and seemingly most inhospitable environments in Britain.

Goats kept for cashmere, therefore, would need even less shelter than other breeds and field shelters would only be provided in the harshest unsheltered environments.

Field Shelters

With any temporary shelter it must be remembered that goats are inquisitive, active and sometimes destructive animals and, therefore, lightweight structures will not last long.

Goats love to jump up onto anything they can so low shelters must be

Figure 3.8   Low shelters must be able to withstand being jumped on.

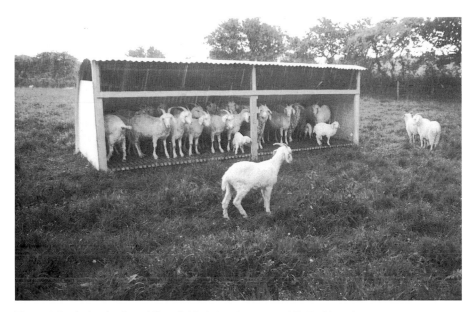

Figure 3.9 A simple slatted-floor field shelter (courtesy of D. Parkinson).

strong enough to withstand this. They also chew anything that takes their fancy and any soft or loose part of a shelter will be vulnerable.

Field shelters need not be very elaborate and should be light enough to be moved at least by towing with a vehicle. Goats sheltering during wet weather soon create a muddy area around shelters.

To prevent the goats having to lie in mud, a slatted floor is an advantage. This is particularly so for milking animals, as udders should preferably remain clean, and for Angoras to avoid soiling of the fleece. Slatted floors can be made from slats 25–35 mm wide with a gap between the slats of about 16 mm. Larger gaps have been used for goats in indoor housing but for the relatively small fibre breeds the gap should not be too wide.

To avoid one bossy goat taking over an entire shelter very wide openings are desirable or alternatively some shelters have been designed with more than one entrance. Figure 3.9 shows an example of a simple field shelter that could easily be constructed on the farm.

Fencing

This is one of the aspects of goat keeping that has given goats the reputation of being difficult animals to manage in a farm environment. According to Mackenzie in his book *Goat Husbandry*, 'The hard, ugly glint that appears in the eyes of some farmers when goats are mentioned is often due to the pleasure that wire netting provides.' He gives an amusing description of the goats' habit of destroying fencing by rubbing up against it.

There are several reasons why goats are difficult to contain by fencing. They are very inquisitive animals and when put into a new field or paddock they will roam around investigating fences and gates. If there are shrubs, bushes or trees close to the fence they will clamber up resting their front feet on the fence in an effort to reach over. These activities soon reveal any weak points in the fence, which, once found, will enable the goats to get out of the field in no time at all. Even hungry goats will spend much time investigating their surroundings before settling down to graze. This behaviour is very different from sheep whose main purpose in life seems to be to feed, grow and be productive.

Well-constructed deer fencing or high chain link supported by concrete posts will remain goat proof for a long time provided it is regularly checked for weak points. The goats' clambering activity will cause stretching, staples to loosen or straining wires to break—particularly as they tend to clamber in the same place in an effort to reach something of interest such as an overhanging branch. For this reason barbed wire should not be used as clambering goats may severely injure themselves.

It must also be recognised that some goats will jump fences. Feral types are probably the most agile and to contain them these fences should be at least 1.5 m high. Some dairy goats will jump although generally not as well as ferals. Anything that may act as a 'launching pad' should be avoided. This would include straining posts, bales of straw, humps in the ground, wheels of parked vehicles or trailers, in fact any horizontal or near horizontal surface of 6 sq cm or more!

In recent years there have been great advances in power-fence technology and many of the problems just described can be overcome with electrified or part electrified or power fencing. Sheep fencing which is not in a very good condition can be made goat-proof by running a single 'hot' wire about 20 cm away from the fence and about the same distance off the ground. Such a wire can be supported on out-rigger brackets attached to the main fence posts (Figure 3.10).

If clambering to reach hedges or overhanging trees is a problem this can be prevented by running a 'hot' wire along the top of the permanent fence.

Very effective goat-proof fencing can be made from a power fence with 4 or 5 galvanised wires (Figure 3.11). Non-conductive timber is available for posts thus doing away with the need for insulators, or, alternatively, some manufacturers supply glass fibre posts.

Power-fencing that is erected and maintained well is an inexpensive way of containing goats. At the time of writing some 5 wire fences cost less than 50p/m. Goats do, however, behave differently to sheep and it may be unwise to rely on power fencing alone as a perimeter fence.

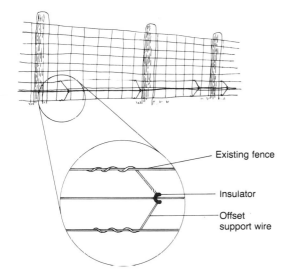

Existing fence

Insulator

Offset
support wire

Figure 3.10   An electrified outrigger
wire or offset on an existing stock-
mesh fence.

Unlike sheep, goats will graze right up to an electrified wire seeming to
prefer the uncontaminated grass and herbage they find there. This may be
the reason why they seem to discover quickly when a fence is not working
and then they may well break out. It is imperative, therefore, that the
fence is regularly tested to see if it is functioning correctly.

Electrified plastic sheep netting can be used as a temporary fence but it

Figure 3.11   A Gallagher five-wire power fence using insultimber posts.

does not deliver a very strong current and with most types only the horizontal strands are electrified. There have been many instances where goats have discovered this and have chewed through the vertical strands!

Electrified netting is not recommended for horned goats as they may get caught when grazing close to the fence and deaths have occurred to both sheep and goats in this way.

Tethering

The problems and expense of fencing can be overcome by tethering but this is not practical on a farm scale and is, therefore, only appropriate to the person keeping one or two house or pet goats.

If goats are tethered it must be remembered that they will not be able to get away from danger such as dogs or maybe children throwing stones nor will they be able to shelter from the extremes of weather. They should, therefore, always be tethered within sight of someone who will come to their rescue when necessary.

Because goats are such active animals it is important to design the tether so that it cannot become tangled in any way. To achieve this it is necessary to have swivels at each end and one in the middle. It is also important for the goat not to be able to get near anything on which it may become caught up. This would include trees, bushes, buildings, farm implements, etc.

Chapter 4

Feeding and Nutrition

The Ruminant System

Ruminants are specialised feeders that, in the wild, fit into a particular niche in the food chain. They have evolved to be able to eat a wide range of plant material that is indigestible to most other groups.

The most striking feature of this group is the modified stomach system that enables them to digest large quantities of plant material. The main organ that facilitates this is the rumen.

A ruminant such as a goat quickly eats plant material and swallows it and later on, usually during a quiet part of the day or night, this material will be regurgitated and thoroughly chewed and then swallowed again. The regurgitated bolus of food material is called cud and the act of chewing it is called cudding or simply chewing the cud.

One theory explaining how this mechanism may have evolved suggests

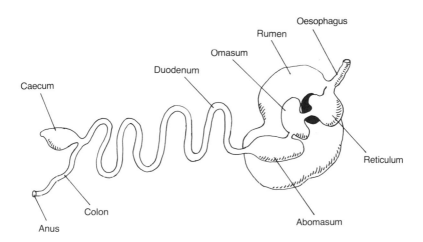

Figure 4.1   Simplified diagram of goat digestive system.

that this enables a group that needs to spend a lot of time gathering food material of relatively low nutritional value to do so quickly. Whilst feeding with their heads down herbivores are vulnerable to attack from predators. Chewing the cud allows them to gather food quickly and to then chew it properly at some later period when they can watch out for predators or when predators will not be active.

The rumen contains a diverse population of micro-organisms that secrete enzymes which will break down the cellulose plant walls thus making the nutrients therein available. The types of organism that make up the gut flora vary according to the type of material eaten. This is the main reason why it may be harmful to make sudden changes to a ruminant's diet. Feed material will be retained in the rumen until it has been broken down to particle size that will pass on through the gut system.

In young ruminants the rumen does not begin to function until they are 2–3 weeks of age. The milk that the young drink bypasses the rumen by way of the oesophageal groove and is digested in the last stomach, the abomasum, which is analagous with the stomach of monogastric animals. If milk does get into the rumen of these youngsters bloat will often occur. This can sometimes happen if they drink milk too quickly.

The starch, fibre and sugar in the food material are broken down by the rumen microbes to release volatile fatty acids (VFA) and gas (methane). The VFAs are absorbed through the rumen wall and are the animal's main source of energy. The proportions of the different VFAs vary according to the food eaten. Acetic and proprionic are produced in greatest quantities but a little butyric acid is also produced.

The methane is removed by eructation (belching) but sometimes this is not possible due to obstructions or torsion of the gut in which case the animal will become bloated. Sometimes the blockage can be due to froth in which case medication can be given to disperse it.

When fibre is fermented large amounts of acetic acid are produced whereas the starch and sugars from cereal-based concentrate feeds produce large amounts of proprionic acid.

The varying amounts and the absorption of the different VFAs will influence the composition of the milk. Acetic acid is used for synthesising milk fat whereas milk yield is governed by lactose production from glucose which itself is synthesised from proprionic acid.

It can be seen, therefore, that the type of material eaten by a ruminant will influence milk composition. This is an important factor for those who produce milk which is used for the manufacture of products such as cheese or yoghourt.

Some of the protein in the feed will be used (degraded) by the microbes in the rumen and some will be passed through the gut to be used by the animal (undegraded). The microbes will break the dietary protein down to make their own protein and some of this microbial protein will be an important source of nutrients for the ruminant animal.

Urea can be fed as a source of non-protein nitrogen which will be utilised by the rumen microbes in addition to any urea recycled from the dietary protein. When urea is broken down in the rumen ammonia is formed and excessive levels of ammonia in the blood will be toxic.

Once the food material reaches the abomasum digestion follows a similar pattern to monogastric animals with absorption of different nutrients taking place throughout the stomach (abomasum) and the large and small intestines. However, food material that has passed through the rumen will be profoundly changed. Much of the protein will be microbial protein, the fibre will be largely indigestible and most of the starch and sugar will have been utilised in the rumen.

FEED AND NUTRIENT REQUIREMENTS

When considering the dietary needs of an animal, particularly a ruminant, it is normal to categorise the animal according to its physiological or productive state. If an animal is full grown and it is not pregnant or lactating it will require only enough nutrients to maintain bodily function or in other words to stay alive and this requirement will be termed the 'Maintenance Requirement'. Some goats, particularly Angoras, will always require more than simply a maintenance ration because they will always be in a state of production. The 'Production Requirement' can be sub-divided into the needs to support pregnancy and lactation. Young goats will have a 'Growth Requirement'.

One of the most important aspects that influences a goat's ability to take in enough nutrients is its appetite. All animals are limited to how much food they can consume and in the case of ruminants this is termed the 'Dry Matter Intake' (DMI). It is important to talk about food material in terms of dry weight as it is that part of the material that contains the nutrients. The fresh weight of something like a root crop, such as turnips, would be very misleading.

The dry matter intake of a goat is about 3.5–5 per cent of its body weight per day. This can be increased if a goat is fed a little food at frequent intervals, particularly if each feed is different and is interesting to the goat. Such a system would be too time consuming and costly for an enterprise but is used by those with small numbers of carefully bred goats to enable their full genetic potential for milk production to be used.

There is some evidence to suggest that the rumen of a goat is proportionally larger than in other ruminants and this possibly gives the goat the ability to utilise the large quantities of food, particularly stemmy forage, that it does.

Some evidence also suggests that goats digest forage more efficiently than other ruminants. However, when fed high-quality forage they seem to be no more efficient than sheep or cows. What a goat does seem able to do more efficiently is to select the more digestible parts of a given forage. Therefore, they make better use of forage of low overall nutritional value by selecting out these more digestible parts.

Nutrient Requirements

Having established that a goat has a limited capacity for food material and that it has different requirements according to its physiological state it is now appropriate to establish what its requirements for specific nutrients are.

Energy

These days energy is measured in joules. Older readers may be more used to calories. One calorie is equivalent to 4.19 joules. Large units of energy are required by animals and the unit megajoule or MJ (1 million joules) is used.

A goat's requirement for energy is usually referred to as its metabolisable energy requirement (ME) and thus it will require a certain number of MJ ME per day. The amount required will vary according to the needs for maintenance, production or growth.

The daily energy requirements for dairy goats are:

Maintenance: 0.5 MJ ME/kg of metabolic body weight (live weight in kg to the power of 0.75)

Pregnancy: 0.5 MJ ME/$kg^{0.75}$ rising to 0.7 MJ ME/$kg^{0.75}$ for the last month

Lactation: maintenance needs + 5 MJ per kg of milk produced

These requirements when properly scaled are virtually identical to those for a British Friesian cow.

Various factors will affect the goat's energy requirement such as activity and the environmental temperature and, in the case of milkers, the fat content of the milk. An extra 1.5–2.0 MJ would be required for goats grazing outside. The actual energy requirements for milking and growing goats are shown in Tables 4.1 and 4.2.

Not enough is known yet about the energy requirements of fibre-producing goats. The Angora is unusual in continuing to grow hair at the expense of body function. Angoras kept in sub-optimal environments, such as the thorny scrubland of South Africa and the USA, have a high incidence of abortion which is now understood to be the consequence of an inadequate energy intake.[1] With sheep, lambs would be produced at the expense of body condition and certainly wool growth.

Protein
Protein is required to build new tissues for growth or replacement in an animal body. It is also necessary for milk and hair production. Also, as

Table 4.1 Nutrient requirements for lactating goats[2]

Liveweight (kg)	Milk yield (kg/day, 3.5% BF)	Dry matter appetite (kg/day)	ME (MJ/d)	DCP (g/d)
	0	1.5	8.0	51
	1	1.7	13.1	99
50	2	1.9	18.2	147
	3	2.1	23.3	195
	4	2.3	28.4	243
	0	1.8	9.2	59
	1	2.0	14.3	107
	2	2.2	19.4	155
60	3	2.4	24.5	203
	4	2.6	29.6	251
	5	2.8	34.7	299
	0	2.1	10.3	66
	1	2.3	15.4	114
	2	2.5	20.5	162
70	3	2.7	25.6	210
	4	2.9	30.7	258
	5	3.1	35.8	306
	6	3.3	41.1	354

Table 4.2 Nutrient requirements of growing goats (per day)[3]

Liveweight (kg)	Dry matter intake (kg)	Liveweight gain (g)	Energy requirement (ME MJ)	DCP (g)
10	0.45	100	6.0	55
		200	9.0	75
30	1.3	100	9.8	70
		200	12.8	90
50	2.1	100	13.0	81
		200	16.0	101

described earlier in this chapter, the rumen microbes need protein to enable them to play their part in the digestive process.

Protein requirement is generally related to that for energy. As the requirement and utilisation of dietary protein by the gut microbes is increasingly understood it is normal now to describe the protein value of a feedstuff by the digestible crude protein content (DCP), and by the proportion degraded and undegraded in the rumen (RDP and UDP). Approximately 9 g of rumen degradable protein should be supplied per MJ ME.

High yielding dairy goats and growing kids may need amino acids (the components of protein) in greater quantities than can be obtained from the microbial protein and, therefore, these must be provided by extra undegradable protein (UDP).

Minerals, Trace Elements and Vitamins

Very little experimental work has been carried out to determine the goat's requirement for minerals and trace elements and as a result there is a shortage of published data. From field observations based on experience it would seem that, if given the opportunity, the goat is able to manage its own mineral metabolism very well.

This is particularly the case when the goats have access to grazing or browse. Goats will eat a much greater range of plant species than either cows or sheep and many of these will be what are normally considered as weed species including deep rooted plants which in some cases are rich in minerals and trace elements. In a free grazing situation, where there is a wide range of plant species, mineral deficiency problems in goats are rare.

The two common problems of sheep and cattle, hypocalcaemia (milk fever) and hypomagnesaemia (grass staggers) are much less of a problem in the goat and yet this small framed high yielding animal would be expected to suffer from these conditions. They would perhaps be expected to be more common in zero-grazed herds fed cut green grass from a sown monoculture ley.

Copper metabolism in dairy goats is similar to that in cows and, therefore, it is not necessary to worry about the copper toxicity problems seen in sheep. There are suggestions that Angora goats may be different from dairy breeds in this respect.[4] Until nutrition scientists produce advice to the contrary, it would be advisable to avoid excessive copper intakes in this breed.

Table 4.3 gives suggested recommendations for mineral intakes which become more important if the goat's natural foraging activity is suppressed. In most cases it is better to supply an excess of minerals and trace elements rather than too little.

Mineral blocks or licks are useful in providing goats with a degree of

Table 4.3 Recommended daily allowances for minerals and trace elements

	Growth	Lactation
Calcium	0.5 g/kg body wt +1 g/100 g gain	18–21 g
Phosphorus		15 g
Magnesium	0.8 g/kgDM	2.5 g/kgDM
NaCl (salt)	0.5% of daily ration	
Potassium	5 g/kgDM	8 g/kgDM
Sulphur (more important in fibre goats)	0.16–0.32% of daily ration	
Selenium	0.1–0.2 mg/kgDM	
Iodine (higher levels recommended if goitregenic feeds such as kale, clover, cabbage, etc are offered)	0.5–2.0 mg/kgDM	
Iron	50–100 mg/kgDM	
Copper (dependent on molybdinum levels)	10 mg/kgDM	
Zinc (high levels of calcium interfere with zinc absorption)	10–40 mg/kgDM	
Manganese	20 mg/kgDM	
Cobalt	0.1 mg/kgDM	

choice and like most animals goats will make use of such supplements when they feel the need. In the writer's experience goats fed a concentrate ration containing a dairy cow mineral supplement and offered mineral licks or blocks in their pens or paddocks are most unlikely to suffer from mineral-deficiency problems.

The minerals that may cause problems if fed in excess are phosphorus and selenium. The ideal ratio of calcium to phosphorus is 2:1 – if the phosphorus level rises above this problems with urinary calculii (urilothiasis) may be induced, particularly in castrated males being reared for meat.

An excess of selenium may cause soreness and sloughing of the hooves, loss of hair (alopecia), reproductive inefficiency, diarrhoea, loss of appetite (anorexia) and a general emaciation.

It is important to realise that these symptoms are common to a number of conditions and, therefore, goats showing any of these should not necessarily be suspected of suffering from mineral toxicity.

Once again it is stressed that goats with access to mixed grazing and/or a reasonable level of mineral supplementation are unlikely to have any problems at all and are almost certainly less of a problem than sheep or cattle.

Details of health problems associated with mineral imbalance are given in Chapter 7.

Water

A goat can drink from 4–18 l (1–4 gal) of water a day depending on climatic conditions, type of feed available and on the physiological state of the goat. A lactating goat requires 1.43 l of water per 1 kg of milk and milk production will be inhibited if water is limited.

Normally goats are fairly fussy about the water they will drink and prefer it to be clean. They can often be encouraged to drink if the water is warm and in fact some goatkeepers always offer warm water to high yielding milkers. The economics of this practice are doubtful as any increase in yield is unlikely to offset the cost of heating the water.

Some tropical breeds are able to go long periods without drinking. The black Bedouin goat is able to store water in its rumen equal to 40 per cent of its body weight and in doing so is able to withstand water deprivation for 2–4 days and yet is able to produce a satisfactory milk yield during that time.[5]

Details of methods of offering water to goats are given in Chapter 3.

FEED MATERIALS

Goats, like all ruminants, are forage eaters with a gut system particularly adapted to a large intake of plant material. In practice this material will be made up of fresh grazing or browsing, cut fresh material carried to the goats or conserved forage material usually fed when fresh forage is not available.

If given a free choice goats prefer browse species, i.e. shrubs, bushes and weeds, rather than grass or clover. However, if the choice is limited they will eat most material considered as typical ruminant feed.

Of the fresh, sown materials goats prefer lucerne and will eat approximately 100 g/kg of metabolic body weight (live weight$^{0.75}$). Rye grass, maize, and vetches will also be readily consumed, 80 g/live weight$^{0.75}$, but some grasses such as fescue are not so readily consumed. Several studies have shown that goats do not readily graze white clover showing some preference for the red variety.

Goats prefer some plant material after it has wilted. One example is the stinging nettle which may not be eaten when growing but seems to be a favourite once cut.

A wide variety of conserved forages are available for feeding to goats when fresh forage is unavailable or for convenience when goats are permanently housed.

The most common conserved forage in temperate climates is hay. A variety of types can be fed, with the highest intakes being achieved from lucerne and red clover hay. These are not readily available everywhere and may be expensive and, therefore, grass hay is more commonly fed. Even with grass hay goats prefer stemmy leafy types and will usually eat seed hay in preference to fine soft meadow hays.

Goats are very selective and can be wasteful forage feeders if they have the opportunity to sort through material such as hay. In selecting the more digestible and palatable parts a lot will be pulled out of hay racks onto the ground and wasted. Feeding systems which reduce waste are discussed in Chapter 3.

All forage must be well made as goats will reject anything that is mouldy or contaminated in any other way. Contrary to popular belief they are fastidious feeders and certainly will not eat everything and anything.

A range of dried forage pellets or cubes such as lucerne or grass are now available and will be eaten as readily as hay. The heating necessary in the drying and cubing process improves the digestibility of the nutrients, particularly the protein, allowing the manufacturers to produce cubes of different nutritive value. These products offer the opportunity to increase the amount of undegraded protein in the ration along with an increase in digestible fibre.

Silage is being increasingly used for goats as more established farms, where silage is made for cattle, are developing goat enterprises. Silage is a more variable material than many other conserved forages and differences in quality and palatability will be reflected in the goat's intake which may range from 50–85 g/live weight$^{0.75}$. Trials at the National Institute for Research in Dairying (now the Institute of Grassland and Animal Production) showed that there was no production advantage from grass silage, compared with medium quality grass hay when fed to lactating British Saanens.[6] A group fed equal quantities of hay and silage did show a small but significant increase in milk yield.

Health problems associated with silage feeding are more common when maize rather than grass silage is used. These include acidosis, cortical necrosis, listeriosis and enterotoxaemia. Listeriosis is more of a problem when bale or bag silage is fed. These and other aspects of nutrition related disorders are described in Chapter 7.

The advent of milk quotas among European dairy cow farms has changed some of the strategies for feeding milking cows. There has been

Figure 4.2 British Saanen goats feeding from *ad-lib* silage at the Institute of Grassland and Animal Production, Shinfield.

an increasing interest in using byproduct feeds of relatively low nutritional value but which are economical to feed when absolute yield is not necessarily the most important parameter. Goats, by nature of their wide choice of food plants and preference for stemmy plant materials with high dry matter content, are likely candidates for some byproduct feed materials. Table 4.7 lists some of these but it is likely that other materials could be added to the list by farmers applying a little trial and error.

Sugar-beet pulp was one of the first of these byproducts to be extensively used as a ruminant feed. It is still extensively used and is now regarded as a feedstuff in its own right. It provides a useful source of digestible fibre with a reasonable amount of protein and energy and it appears to be very palatable to goats.

Brewers' grains are another byproduct feed that has been used for ruminant feeding for a long time. Again it provides digestible fibre, it is palatable and provides more protein but less energy than beet pulp.

Traditionally amateur goatkeepers have, quite naturally, used a wide range of garden and kitchen waste to help reduce the feed bill for their goats. Some of these materials are available in large enough quantities to be considered by the larger-scale commercial goat farmer.

Brassicas such as kale, rape and cabbage may be available and are a useful supplement to the diet as they are a relatively good source of energy

Table 4.4 Some common poisonous plants

Acorns (unripe)	Deadly nightshade	Parsnip*
Anemone	Dog's mercury	Privet
Beet leaves*	Foxglove	Prunus leaves (when dry)
Bindweed	Henbane	Ragwort
Box	Kale	Rhododendron
Bracken*	Laburnum	Rhubarb leaves
Bryony	Laurel	Snowdrop
Buttercup	Narcissus	Spindle
Celandine	Oak leaves*	Yew
Charlock (seeds)	Potato (raw)	
Daffodil	Potato haulm	

NB Most evergreens are toxic to some degree
* May taint milk

and protein. However, these materials can interfere with iodine absorption and may cause enlargement of the thyroid gland (goitre). Large quantities of kale may also cause anaemia. It is suggested that these materials should not make up more than 30 per cent of the dry matter intake (DMI) of lactating goats.

Root crops such as carrots, turnips, swedes and artichokes can all be fed but the goat's intake of these will be influenced very much by the condition in which they are offered. Whole roots covered in soil will not be eaten as well as those that are clean and chopped. Roots do have the advantage of being relatively easy to store and, therefore, can be a useful supplement during the periods when fresh food may be unavailable.

High yielding dairy goats will not be able to produce their full potential from forage or byproducts alone and, therefore, it will be necessary at times of peak yield or high demand to provide food that contains nutrients in a more concentrated form. It is not surprising that these feeds are usually referred to as concentrates.

Concentrate feeds include all of the cereal grains such as barley, wheat and oats, legume seeds such as peas and a variety of beans and a range of extracted meals such as cotton cake, rapeseed, soyabean, and linseed. Many of these are heated during processing and this can protect the protein from rumen degradation. Fish meal is another concentrated source of protein of which a large proportion will be undegraded.

Goatkeepers tend to favour oats as a major ingredient in concentrate rations but all cereal grains are suitable for goats. These days barley is used in many dairy cow rations which can satisfactorily be fed to goats. Very high levels of cereal grains, particularly flaked maize, can induce conditions such as acidosis.

Poisonous Plants

Adult goats in a natural grazing or browsing situation are rarely poisoned. They seem to appreciate which plants are harmful. However, in an artificial situation where food is being offered to penned goats they may eat harmful material out of curiosity or they may eat enough mildly toxic material to induce illness. Table 4.4 is not definitive but does list most of the common toxic plants.

THE NUTRITIVE VALUE OF FEEDSTUFFS

As may be understood from the information so far, a bewildering range of materials are available as possible feedstuffs for goats. The decision on whether or not to use them or which combination to use will depend on a number of factors.

Probably the most obvious question to be asked is will goats eat it? Care should be taken with the answer because goats, being inquisitive and yet fastidious feeders, may eat something out of interest but not on a regular basis or alternatively they may initially reject something which, given time, they will accept and readily eat.

Generally speaking goats will eat a wide range of feedstuffs as long as they are presented in good condition and have not become spoiled or contaminated. Once it has been established that the goats will eat the feed the next consideration is how much will they eat?

It is the answer to this question which relates directly to the nutritional value of the feed. If 1 kg is the maximum amount of a given material that a goat will eat and that 1 kg contains half of the daily energy requirement of the goat it will not be able to perform to its full potential. A way has to be found to increase its intake of the material or a more energy dense feed must be provided.

When assessing the value of a feed material both in monetary and nutritional terms it is important to know the dry matter content of the feed. From Tables 4.5 and 4.6 it can be seen that the amount of metabolisable energy (ME) in turnips is approximately the same as in barley. However, turnips are only 10 per cent dry matter and the rest is water, compared with 86 per cent DM in the barley. It can be calculated from this that 1 kg of barley would have the same amount of ME as 8.7 kg of turnips.

This has two important implications. First of all it may not be possible for a goat to eat enough turnips to satisfy its energy needs. Secondly, although the price of a feed material may appear value for money, when the feed value is taken into account it may turn out to be very expensive indeed.

Table 4.5 Nutritive values of some forages and rootcrops for goats

	Dry matter (%)	ME (MJ/kgDM)	Crude protein (g/kgDM)	RDP (g/kgDM)	UDP (g/kgDM)	Crude fibre (g/kgDM)
Silage						
Grass						
(good quality)	27	10.2	170	136	34	300
Lucerne	25	8.5	168	101	67	296
Maize	21	10.8	110	66	44	233
Red Clover	22	8.8	205	123	82	300
Hay						
Grass						
(good quality)	85	9.0	101	81	20	320
Lucerne						
(half flower)	85	8.2	225	180	45	302
Red clover	85	8.9	161	129	32	287
Straw						
Spring barley	86	7.3	38	30	8	394
Spring oats	86	6.7	34	27	7	394
Winter wheat	86	5.7	24	19	5	426
Roots						
Potatoes	23	12.5	90	72	18	38
Turnips	9	11.2	122	98	24	111
Swedes	12	12.8	108	86	22	100
Greens						
Grass						
(monthly cut)	20	11.2	175	105	70	225
Lucerne (in bud)	22	9.4	205	123	82	282
Cabbage	11	10.4	136	82	54	182
Kale	16	11.1	137	82	55	200

Other factors which will also determine the relative monetary value of feed materials are palatability, cost of transport, ease of storage, ease of handling, etc. It will often be the case that, when all these things have been considered, barley bought in bulk will actually be better value or cheaper than some feed materials that may be on offer for half the price of barley.

The nutritive value of a feed material will depend on several parameters:

Digestibility

Usually expressed as the D-value or the Digestible Organic Matter Content (DOMD). This is the percentage of organic or non-mineral

Table 4.6   Nutritive values of grains and other seeds

	Dry matter (%)	ME (g/kgDM)	Crude protein (g/kgDM)	RDP (g/kgDM)	UDP (g/kgDM)	Crude fibre (g/kgDM)
Barley	86	13.7	108	86	22	53
Oats	86	11.5	109	87	22	121
Wheat	86	14.0	124	99	25	26
Maize	86	14.2	98	59	39	24
Field Beans	86	12.8	290	232	58	85
Peas	86	13.4	262	157	105	64

Table 4.7   Nutritive values of some byproduct feeds

	Dry matter (%)	ME (g/kgDM)	Crude protein (g/kgDM)	RDP (g/kgDM)	UDP (g/kgDM)	Crude fibre (g/kgDM)
Dried brewers' grains	90	10.3	204	122	82	169
Sugar beet pulp (dried & molassed)	90	12.2	106	64	42	144
Maize gluten	90	14.2	394	315	79	23
Wheat middlings	88	11.9	176	141	35	86
Wheat bran	88	10.1	170	136	34	114
Linseed cake	90	13.4	332	199	133	102
Meat and bone meal	90	7.9	527	211	316	0

matter, in the dry feed material, that is digested by the animal. The values may range from 40–90 per cent.

Metabolisable Energy (ME)
This is a measure of the amount of energy in a feed material that becomes available to the animal following digestion and is vital to maintain body function and for the needs of growth and production of milk and fibre. The unit of ME is the megajoule (MJ or 1 million joules). For those who are more familiar with the old unit for energy, the calorie, 1 calorie is equal to 4.184 joules. The ME value of a feed material is normally expressed as MJ ME per kg of dry matter (DM).

Digestible Crude Protein (DCP)
This gives an estimate of the protein available to the animal. It is usually expressed as grams per kilogram of dry matter (g/kgDM). It is now realised that the efficiency of protein utilisation is influenced by the degree

of microbial degradation of the protein in the rumen. Only a relatively small amount of protein will be available to the animal if a large proportion is degraded in the rumen.

In modern tables of feed values some indication of the amount of protein degraded in the rumen (RDP) and the amount undegraded (UDP) will be given. All of these values can only be a guide as samples of the same type of feed will vary according to time of harvest, climatic conditions and many other factors. Anyone looking for ingredients of a particular quality is advised to have a sample analysed.

Making Up Rations For Goats

The choice of ingredients when formulating a ration will depend on several parameters and the order of importance of these will be different depending on the interests and circumstances on each farm. All of the following points would be considered:

1. Nutritive value.
2. Availability of the ingredients.
3. Cost with respect to nutritive value.
4. Palatability, will the goats eat it?
5. Storage facilities.
6. Availability of natural grazing or browsing.
7. Approximate weight and production performance of the goats.

Generally speaking farmers will want to compromise between cost, convenience and performance. This may be different for the breeder–enthusiast who may be most interested in performance with less emphasis on cost. This is quite appropriate when dealing with very small numbers of goats involving a relatively small feed bill. The difference between £140 and £180/tonne will hardly be noticed on a small unit whereas on a large farm with a large annual feed bill it would make a very large difference.

Time is likely to be more critical on a farm and, therefore, rations and feeding systems will need to be fairly straightforward. Those who keep goats for pleasure, quite rightly, enjoy experimenting with different feeds and different ingredients – something a farmer may not have time to do.

There is a tendency amongst hobby goatkeepers to feed goats coarse or loose concentrate mixes where the ingredients are mixed together without any form of pelleting or cubing. Such mixes tend to be expensive if bought from a merchant and time consuming if mixed at home. They are also wasteful because the goats are able to pick out the ingredients they particularly enjoy. It is rare for all goats to eat all ingredients and, therefore, a lot is wasted. Cubed or pelleted feeds overcome this problem,

Table 4.8 Some rations suitable for goats with different nutrient requirements

	Dairy goats						Angora goats	
Liveweight (kg)	70	70	60	60	60	55	50	50
Milk yield (kg)	5	3	4	3	–	2	2	–
State of pregnancy	–	–	–	–	late	–	–	late
				kg/day				
Feedstuffs								
Good hay	1.5	–	1.0	1.2	1.5	1.0	0.75	0.7
Silage (early cut)		7.0						
Kale	2.0							
Grazing						spring 6.0	6.0	4.5
Dried sugar beet pulp	0.5	0.5	1.0	0.4				
Medium energy (12 MJ/kgDM) concentrate	–	0.5	–	1.4	0.3	–	18%CP	0.5
High energy (13 MJ/kgDM) concentrate	1.25		1.0					
% forage in ration	40	67	33	53	83	100	100	80

although it may take some time for goats used to loose mixes to accept cubes.

By considering the points set out earlier in this section it is possible to work out a ration to suit goats in any situation as long as approximate body weights are known. A weighing crate is recommended as a useful management tool not only with respect to feeding but also for checking the health of the goats. Once one has ascertained the weight of the goat or goats and assuming one knows their physiological status it is possible to calculate their nutrient requirements from the information given earlier in this chapter.

Once one has calculated these requirements it is then possible to design a ration based on the feedstuffs and ingredients that are available bearing in mind the parameters already mentioned such as cost, palatability, etc.

As everyone's circumstances are likely to be different it is not appropriate to give recommendations of particular rations but in Table 4.8 some examples are given for goats of different nutritional needs. These examples should be used as guides only for those who want to formulate rations for their goats using whatever feed materials are economic to use in their part of the country.

It is very important to give goats plenty of time to get used to new feedstuffs. For example, they may refuse to eat a new concentrate ration for several days. With dairy goats it is sensible to try such changes when they are dry, or at least towards the end of lactation, to avoid losses in milk production. It is always best, if possible, to introduce new rations to young goats as they tend to be less fussy and will be quite used to them by the time they are in production. Any feed material that is soiled, mouldy or spoilt in any way will be wasted and, therefore, even when available at low cost, will be uneconomic when compared with good quality well-made material.

At the conclusion of this chapter it is important to reiterate that all the data that has been presented should only be used as a guide when working out feeding strategies. It is unlikely that tuning goats' feed rations to the last fraction of a megajoule or to the last few grams of milk or fibre can be justified. In a large herd if the feed offered is adequate to realise the full potential of the average goat and if one constantly strives, by careful selection, to upgrade the average goat in the herd this will be more economic than detailed attention to the needs of each and every individual, particularly bearing in mind some of the idiosyncratic demands of some goats!

REFERENCES

1. Wentzel, D. (1987), 'Effects of nutrition on reproduction in the Angora goat', *Proceedings of the 4th International Conference of Goats* (Brasilia).
2. M.A.F.F. Agricultural Development and Advisory Service (1984), *Feeding Dairy Goats*, booklet produced by ADAS Nutrition Chemistry Section.
3. Wilkinson, J. M. and Stark, Barbara A. (1987), *Commercial Goat Production* (B.S.P. Professional Books, Oxford).
4. Humphries, W. R., Morrice, P. C. and Mitchell, A. N. (1987), 'Copper poisoning in Angora goats', *Veterinary Record*, Vol. 121.
5. Shkolnik, A. and Silanikove, N. (1981). 'Water Economy, Energy Metabolism and Productivity in Desert Ruminants', Vol. 1 of *Proceedings of the International Symposium on Nutrition and Systems of Goat Feeding* (Tours, France).
6. Badamana, M. (1987), Ph.D. thesis, Department of Agriculture, University of Reading.

Chapter 5

Goat Breeding

Before describing the various methods and techniques that are used when breeding goats it is necessary to consider the anatomy and physiology of reproduction. In this respect the goat resembles the sheep and anyone familiar with these will need very little new information to understand this aspect of goat farming.

MALE ANATOMY

The most obvious part of the male's reproductive system is the scrotum containing the testes. This may vary in size according to breed but, in general, abnormally small testes are a sign of likely low fertility. The scrotum not only supports and protects the testes but also it is an important means of temperature regulation. Normal production of spermatozoa occurs at a temperature 4–7°C lower than body temperature. Thus in hot weather the scrotum will allow the testes to hang well down from the wall of the abdomen and conversely when cold they will be drawn up close to the body. In extremely hot weather this complex temperature regulatory mechanism may break down resulting in poor spermatozoa production. In some goats, particularly in some Angoras, the scrotum may be almost completely bifurcated ('split purse'). This is considered a fault and will be marked down in a show animal. It is unlikely, however, that this would cause any problems of fertility except in very extreme environments.

Failure of the testes to descend into the scrotum will also cause problems of fertility. One testis may not descend (monorchid) or in some cases both may not (cryptorchid). When purchasing males for stud it is most important to ensure that both testes are in place in the scrotum and that they are of reasonable size with no abnormal swellings and that they feel firm and not soft and spongy.

The other external sex organ that is obvious in the male is the penis. In the male goat the penis is normally retracted into a tube called the prepuce. To give extra length during copulation the penis has an S-bend known as the sigmoid flexure (Figure 5.1). On the end of the penis is the thin tubular protrusion of the urethra known as the urethral process.

54

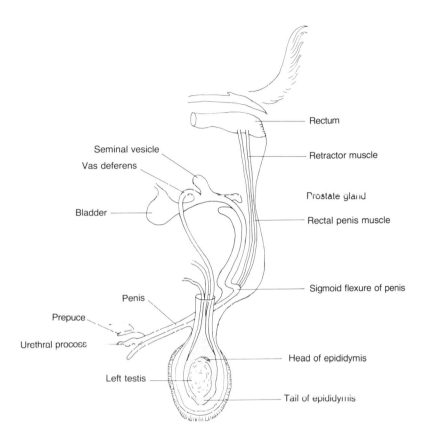

Figure 5.1 Reproductive organs of the male.

When the penis is protruded from the prepuce, particularly during the breeding season, the male goat is able, with remarkable directional accuracy, to spray urine over himself or anyone who is standing close enough!

Physiology

In temperate regions both male and female goats show seasonal sexual activity and in the Northern Hemisphere this usually means sexual activity begins in the autumn. However, if stimulated by receptive females or if they are trained, male goats will mate all the year round. Young male goats are particularly precocious and fertile matings have been recorded from kids of 3 months of age.

Spermatozoa are formed from cells in the testes called spermatogonia. These spermatogonia divide repeatedly to form spermatids which eventually form the spermatozoa which are discharged into the lumen of seminiferous tubules (Figure 5.1). The spermatozoa travel along in fluid secreted by the tubules until they reach the epididymis where they are stored. These newly formed immature spermatozoa are immotile and are very sensitive to unfavourable temperature and nutritional conditions. Full maturation occurs in the tail of the epididymis and the spermatozoa become motile during ejaculation when they come into contact with the secretions of the accessory glands (the vesicular, prostate and bulbo-urethral glands). It takes approximately 50 days from the formation of the spermatozoa in the seminiferous tubules to the time they are stored in the tail of the epididymis. During periods of intense sexual activity this may be reduced as the movement of the spermatozoa through the epididymis may be speeded up.

Another important function of the testes is the production of the hormone testosterone. The secretion of this hormone is controlled by gonadotrophic hormones secreted by the pituitary gland situated at the base of the brain. Although sexual desire in male goats is influenced a great deal by the presence of receptive females, nutritional status and environmental factors also play an important part.

Prior to mating a male goat will spend varying amounts of time in courtship behaviour which almost certainly is important stimulation for both male and female. During hand-mating of pedigree animals, when a female is led to a specific male, it is important that restraints are not imposed on this behaviour and the most successful results will always be from animals that are allowed some time together in a pen or paddock. Normally poor libido is not a problem with male goats.

FEMALE ANATOMY

Unlike the male most of the female's reproductive organs are internal and would only be seen by those attending a post-mortem examination. The only external feature is the vulva which undergoes some changes during oestrus and when kidding (parturition) is imminent.

The vulva opens into the vagina (Figure 5.2), which is where the male's penis deposits semen during copulation. In a normal adult goat the vagina is approximately 8 cm in length. At the end of the vagina is the cervix or neck of the uterus. The cervix varies in length from about 4–8 cm and is made up of 5–6 muscular rings (Figure 5.3) which effectively act as a seal between the vagina and the uterus. The uterus is made up of two large tubes or 'horns' and at the end of each of these horns are the oviducts and

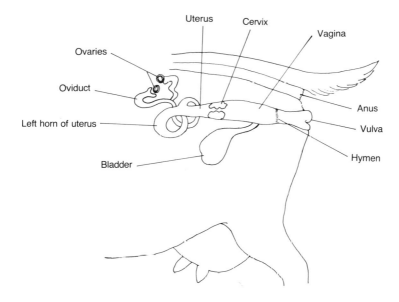

Figure 5.2 Female reproductive organs.

the ovaries. The ovaries change in appearance according to the stage in the reproductive cycle. The eggs or ova are shed from what are called the Graafian follicles and these can be seen during a post-mortem examination if they are near to maturation.

When an ovum is shed the remaining structure is called a corpus luteum, meaning yellow body, and these also can be seen on the ovary and are an indication of a goat that is ovulating normally. If the goat is pregnant the corpus luteum remains and plays a part in maintaining the state of pregnancy. If the goat is not pregnant the corpus luteum regresses.

Physiology

In temperate countries, such as Great Britain, the decreasing daylight after the 21 June, the longest day, triggers off breeding activity in goats. The lengthening nights cause increased release of the hormone melatonin from the pineal gland within the brain. This then causes the release of gonadotrophin releasing hormone which stimulates the pituitary gland into secreting follicle-stimulating hormone (FSH). As its name suggests FSH stimulates the development of the follicle within which an ovum will develop and from which it will be released. The onset of the sequence of events gives rise to oestrous behaviour, or heat, in the goat and the whole cycle of events is called the oestrous cycle (Figure 5.4).

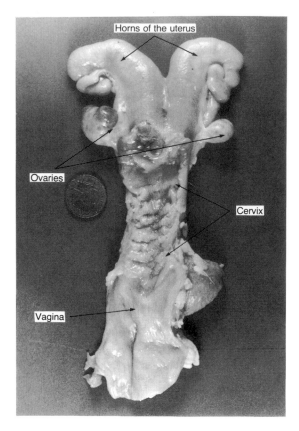

Figure 5.3 A dissected female reproductive tract showing the cervix (courtesy of the Food Research Institute, Reading).

As the Graafian follicle matures it secretes the hormone oestrogen which eventually stimulates the brain into triggering off the release of luteinizing hormone (LH) into the bloodstream. This release of LH causes the follicle to rupture and an ovum will be released into the oviduct.

About 36 hours before ovulation occurs the female goat will normally begin to show oestrous or heat behaviour. This behaviour is a combination of signals to the male that she is at the correct period in her ovulation cycle for mating when changes in the reproductive tract, to facilitate mating, have occurred. The vulva becomes swollen, copious mucus is produced and the cervix dilates.

If a fertile mating takes place the fertilised embryo develops freely in the uterus for about 21 days until implantation takes place and the embryo becomes attached to the wall of the uterus by way of the placenta. The caruncles which form the points of attachment on the uterine wall are present all the time and can even be seen on the uterus of a pseudo-pregnant goat.

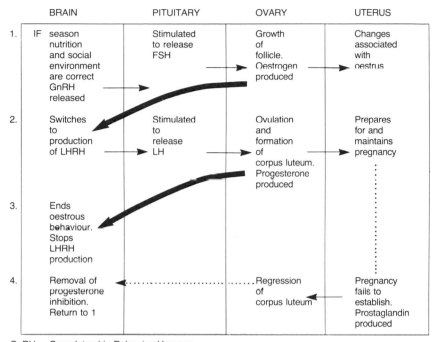

BRAIN	PITUITARY	OVARY	UTERUS
1. IF season nutrition and social environment are correct GnRH released	Stimulated to release FSH	Growth of follicle. Oestrogen produced	Changes associated with oestrus
2. Switches to production of LHRH	Stimulated to release LH	Ovulation and formation of corpus luteum. Progesterone produced	Prepares for and maintains pregnancy
3. Ends oestrous behaviour. Stops LHRH production			
4. Removal of progesterone inhibition. Return to 1		Regression of corpus luteum	Pregnancy fails to establish. Prostaglandin produced

GnRH = Gonadotrophin Releasing Hormone
LHRH = Luteinising Hormone Releasing Hormone
FSH = Follicle Stimulating Hormone
LH = Luteinising Hormone

Figure 5.4 The physiological and behavioural changes that occur during the oestrous cycle.

If a goat is pregnant the corpus luteum, formed after the rupturing of the follicle, remains and produces the hormone progesterone. Progesterone acts as a signal to the brain to shut down the cycling mechanism and prepares for and helps maintain pregnancy. In some animals, but not goats, the role of the corpus luteum is taken over by the placenta.

If conception does not occur the corpus luteum regresses and the level of circulating progesterone consequently falls. The cycle then starts again and a non-pregnant goat will continue to cycle in this way every 21 days until the end of the breeding season, usually during January or February.

SELECTION FOR BREEDING

Most people who farm goats, or for that matter any animal, would expect to gradually improve the productivity of their stock. Big improvements can often be made by changes in husbandry so that the animals become fitter, healthier and better fed. There will be limitations, however, on how much productivity can be increased in this way. These limitations will be the result of the genetic make-up of the animals. In other words all animals are born with a potential for production and that potential will be the result of the mixing of characteristics inherited from the animals' parents, grandparents and, in fact, all of its ancestors.

By selecting animals with certain characteristics and mating them it is possible to gradually improve the performance of that line, generation by generation. Some characteristics are readily passed on and are highly heritable; some are not readily passed on and are referred to as of low heritability.

It is not possible to discuss genetic gain or improvement without a basic understanding of genetics and for the purposes of this book a simplified explanation will be given. All inherited characteristics are carried by genes which occur in pairs, one from the father and one from the mother. A pair or more of genes will control a particular characteristic and thus in the case of colour a goat will either be coloured or white. If the goat has a gene for colour from one parent and one for white from the other it will, in fact, be white because white is what is called a dominant gene and colour is what is called a recessive gene. If two different genes for a particular colour come together the dominant gene will always be expressed. If in our example the goat had received genes for colour from both its parents it would then have been coloured.

If an animal is carrying a pair of identical genes for a particular character, such as the coloured offspring in the example it is known as homozygous for that particular character. If it carries different genes like the goat in our example with the genes for white and colour it is termed heterozygous.

The appearance of an animal as controlled by its genetic make-up is referred to as its phenotype. Our white goat is phenotypically white but its genotype is white/coloured. This is shown in Figure 5.5 where two goats are mated. One is homozygous for white and will be genotypically white. The other is heterozygous and will be phenotypically white because white is dominant.

When these two are mated their kids, or what is called the F_1 generation, could be like the parents either heterozygous or homozygous white. If, however, both parents were heterozygous white as in Figure 5.6 the offspring would be either white or coloured in the ratio 3:1 with 1 homozygous and phenotypically white whereas the other 2 whites would be heterozygous and 1 would be homozygous for colour and would therefore be (phenotypically) coloured. It is rare for a single gene to control a characteristic or trait as shown in the simple example but it serves to show how characters are inherited.

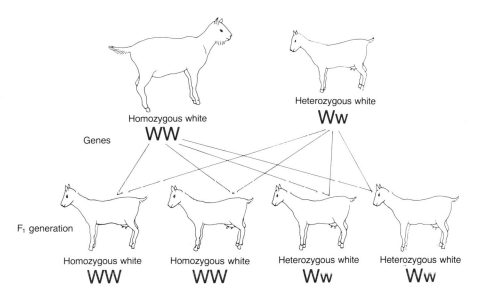

Figure 5.5   A simplified diagram showing the result of crossing two white goats where the male is pure (homozygous) white and the female is carrying the dominant gene for white (W) and the recessive gene for colour (w). All of the progeny are white because of the dominance of the white gene.

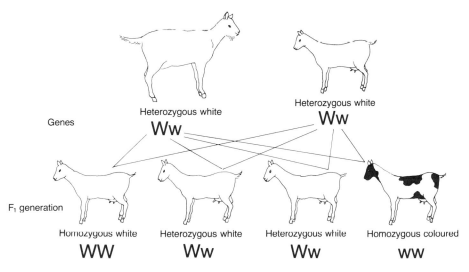

Figure 5.6   A simplified diagram showing the result of crossing two goats both carrying the white (W) and coloured (w) gene. The progeny or F₁ offspring will be white and coloured in the ratio 3:1.

With selective breeding the intention is to cross animals together in such a way that the progeny will hopefully be even better than their parents. It helps if the genetic make-up of the parents is known which may be the case for some characters. Using our example again we know that a coloured goat must be homozygous for the gene for colour. If it wasn't, the dominant gene would cause it to be white.

Some characteristics are linked to others and it may be, in selecting for one desirable feature or trait, that one also selects for an undesirable feature which may cancel out or be even less desirable than the trait that was being selected for.

One such trait in goats is polledness or hornlessness. If a naturally polled male is mated with a female carrying the gene for polledness there is a good chance of producing female offspring that will be homozygous (pure) polled animals. These will be inter-sexed, which means they may have some parts of the male and female reproductive tract and characteristics and they will be infertile. They are not, as some people describe, hermaphrodites which means possessing both female and male sexual organs. Naturally polled males do occur without this problem of inter-sex but evidence suggests that fertility is lower in these.[1]

It can be seen from these simple examples that genetics and inheritance are not simple and it would be all too easy to breed/select for a particular

Table 5.1   Degrees of heritability of some important traits for goat production

Angora Traits	
Total yield	0.48
Fibre length	0.22
Greasy fleece weight	0.15–0.40
Clean fleece weight	0.20
Fibre diameter	0.12
Face cover	0.31–0.59
Kemp score	0.20–0.43
Body weight	0.30–0.50
Weaning weight	0.20–0.55
Dairy Goats	
Annual milk yield	0.36–0.64
Butterfat (%)	0.32–0.62
Protein (%)	0.59
Lactose (%)	0.38
Milking time	0.67
Litter size	0.07–0.24
Birth weight	0.01

NB **Higher values indicate greater heritability.**

trait to find that one had also selected for some undesirable and un-economic trait as well.

If traits or characteristics are of low heritability the genetic gain achieved by selecting specific animals showing those characteristics will be less than for traits of high heritability. Thus the hoped for improvement will be achieved only slowly over a number of generations. Table 5.1 shows various traits that will be of interest to goat farmers and shows their heritability. Those with the highest numbers are those with the higher degree of heritability.

By constantly trying to produce better goats and by only breeding from animals that reach a certain level of performance a farmer should gradually and consistently improve his herd. AI, potentially, gives better access to top quality males and, therefore, could be instrumental in a dramatic improvement of the performance of a herd or population.

GROUP BREEDING SCHEMES

If a group of farmers get together it is possible that they can accelerate the improvement of their stock by organising a group breeding scheme. Figure 5.7 shows diagrammatically what is involved. The scheme is easier to operate with fibre-producing breeds because males can be judged by their fleece quality as well as their general conformation. It could be done with dairy goats but performance would have to be judged retrospectively once a male's daughters were in production. In both cases the scheme would depend on semen being collected and frozen.

From within a group of farms two superior males are chosen. This may be by a combination of conformation and performance which could be judged by the milk yields of his daughters or by his and his progeny's fleece quality in the case of Angoras. Semen is collected from these two males and this is made available to all members of the group for use on half of their females. The other females are mated to their own males on the farm. Once a year on the same date all of the males produced are judged according to the parameters described and if any appear to be better than either of the élite males at the top of the 'pyramid' these become the élite males to be used on half the females and so on. This way better and better males will be used with a consequent gradual improve-ment in all the participating herds. By using farm males on half of the goats on any one farm the chances of a good male appearing from 'among the ranks' is not excluded.

Because the farmed goat population is still small, it is probably prema-ture to think of using such a scheme in Britain with respect to the testing of

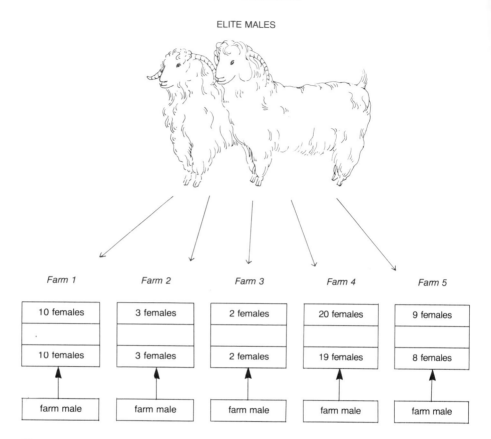

Figure 5.7 A plan of a group breeding scheme showing how two elite males are used on half the goats within the group and males on each farm are used on the rest.

progeny on the basis of milk yield but a scheme could certainly be operated looking at the fleece quality of Angora or cashmere goats.

There has been a tendency, particularly amongst small-scale goat-keepers, to use the most convenient male regardless of his quality and consequently the performance of some populations has gone down rather than improved. Hopefully the increasing use of AI will overcome this problem.

MANAGEMENT OF REPRODUCTION

Having considered the physiological aspects of reproduction we can now consider the practical aspects and examine how we can best make use of our knowledge to maximise the goat's reproductive performance by achieving optimum numbers of healthy kids without undue stress to the mother.

The Breeding Season

In Britain all goats are seasonal breeders and will normally first come into oestrus (heat) during the autumn. The actual season varies a little between breeds and those that originate from tropical countries, such as the Anglo-Nubian, tend to have a longer season than those that originate from temperate countries such as the Swiss breeds.

Usually the season begins in September and the goats will continue to cycle every 21 days if not mated until possibly February. Breeders should not depend too much on getting animals mated at the end of the season as, all too often, the cycle before the one when mating was planned turns out to be the last for that season!

The sexual activity of male goats also tends to be seasonal although they will mate all the year round if regularly presented with oestrous females. Kids show sexual activity earlier than older goats and, therefore, it is unwise to leave male and females together after July.

The onset of the breeding season is influenced by light and to a lesser extent temperature. The increasing length of the night after midsummer causes the release of the hormone melatonin which signals to the brain to begin the cycle of hormonal activity, controlled by the pituitary gland, which results in ovulation in the female.

Control of the Breeding Season

The restriction of seasonal breeding is a problem to some farmers. Seasonal breeding results in seasonal milk production and for those who depend on milk production for their livelihood it is an advantage to be able to produce and supply milk all the year round.

There are two ways of stimulating goats to breed out of season: one involves the administration of hormones or analogues of hormones and the other involves the alteration of environmental conditions, usually light, to induce ovulation.

The most common method of hormone treatment involves the use of sponges impregnated with the hormone progesterone or a synthetic version of it. These sponges are inserted into the vagina and are left to release the hormone over a predetermined period. Intra-vaginal sponges were originally developed for synchronising oestrus in sheep and are still useful for this in goats. Table 5.2 shows the various regimes used according to the time of the year.

Table 5.2 The regime for using progesterone sponges, which are inserted into the vagina for 21 days

Time of year	PMSG Injection	Dose of PMSG
Feb.–June	48 hours before sponge removal	600 IU
July–Aug.	48 hours before sponge removal	500 IU
Sept.–Jan. (breeding season)	At time of sponge removal	400 IU

Nylon threads are attached to the sponge to facilitate removal. It will not be a surprise to those with experience of goats to learn that these threads may be chewed off by other goats and, therefore, some ingenuity may be required when it is time to remove the sponge. Pregnant mares' serum gonadotrophin (PMSG) injections are given as a source of Follicle Stimulating Hormone (FSH) and Luteinizing Hormone (LH) to increase the ovulation rate and to thus increase the chance of conception. The manufacturer's full instructions on how to use intra-vaginal sponges will be in or on the packet. As can be seen the sponge technique can be used both for inducing oestrus in advance of the natural season and for synchronising oestrus within the breeding season.

It is also possible to synchronise oestrus by injections of synthetic prostaglandin, a substance produced by the uterus of non-pregnant goats which causes the degeneration of the corpus luteum. This can only work if there is an active corpus luteum. Goats respond quickly to prostaglandin and usually come into oestrus 24–48 hours after treatment. If a timed mating is required for a goat whose last oestrus is unknown, 2 injections are given 11 days apart.

'Natural' Methods

Various methods of control of the breeding season are used that do not involve hormonal injections or other similar treatment.

Male Effect

If a male is run with females during the transitory period prior to the expected breeding season he will tend to stimulate those females into oestrus some 2–4 weeks early and they will tend to be synchronous. If selected mating of the females is intended then a vasectomised teaser male can be used.

Goat-keepers have, for many years, applied the male effect when their goats failed to exhibit oestrus. In this case a 'billy rag' is used. This is a rag that has been rubbed over the male thus becoming impregnated with his very characteristic smell. This rag will be kept in a screw-top jar and will be brought out to be waved under the nose of a female that is not showing any signs of coming into oestrus when it is thought she should. Often this will be enough to start a female cycling.

Light Effect
Having discussed earlier in the chapter how differences in day length are the main trigger for the onset of the breeding season it is perhaps not surprising that this sequence of events can be mimicked using artificial lighting regimes.

Some goat-keepers have sometimes, unwittingly, used this technique when housing goats in late summer or early autumn. If the shed is fairly dark it is possible that this will accelerate the shortening day effect and the goats may well show first oestrus a few weeks earlier.

The role of the hormone melatonin in controlling seasonal oestrus behaviour was discussed earlier in the chapter. Research work has been carried out, particularly on sheep and deer, to investigate using the hormone to 'override' or minimise the effect of light. By administering melatonin, which can be done in the feed, scientists have been able to induce ovulation and oestrous behaviour out of season. If such techniques could be used with goats this would be extremely useful for a farm wishing to produce milk throughout the year.

Work, particularly in the USA, has shown that out of season mating can be planned by using a controlled artificial lighting regime. Various regimes can be used all working on the principle of a period of long artificial days followed by a period of shorter days. Ashbrook described a system involving 60 days of 20 hours light, during January and February, followed by ambient lighting from 1 March. This resulted in the goats showing a single oestrous period during late April through to June with most showing oestrus in May.[2]

The level of light recommended by Ashbrook was 1 ft of 40 watt fluorescent tube per 10.5 sq ft of floor space with the tubes approximately 9 ft above the floor. Those goats not mated went on to cycle normally in the autumn. There therefore would be no need to separate those goats to be mated during the artificial season from those to be mated during the normal season. The increased lighting is also likely to increase winter feeding activity which will probably increase milk yields.

Various farmers in Britain have used similar techniques and have found that one good halogen vapour lamp either end of a large goat shed or barn is adequate. Ashbrook suggested that the males should be kept in the same extended light conditions if they were to work satisfactorily out of season. Some farmers in Britain have found that this need not be the case and that males put in with the females a few weeks before the artificial oestrus is expected will work satisfactorily.[3] Some have even hand-mated success-fully and have achieved good conception rates even though it has been shown that the 'standing heat' period is shorter in the light treated goats.

Oestrous Detection

Unlike cows, oestrous or heat detection in goats is not normally a problem. Goats exhibit a number of behavioural signs which in those that are regularly handled, such as milkers, are easily recognised.

The most recognisable sign of oestrus is the plaintive cry that nearly all goats will make at this time. It is quite different to their normal call and will soon be recognised once one is familiar with the normal calls and behaviour.

If there are males on the farm the plaintive crying will be accompanied with wistful looks towards the males and if they get the opportunity, the females will stand around the males' pens showing what in human terms would be called flirting behaviour!

The females usually wag their tails rapidly from side to side at this time and sometimes a clear discharge may be seen coming from the vulva. The vulva may also be slightly swollen and reddened.

Figure 5.8   Raddle harness on a 'teaser' male (courtesy of the Food Research Institute, Reading).

Female goats usually go off their food during oestrus and the yields of milking goats will be reduced. Oestrus lasts for about 3 days with ovulation occurring about 36 hours after the onset.

More rarely than in cattle, female goats in oestrus will mount each other. Usually the oestrous goat stands to allow another to mount her.

In spite of all these signs oestrus is sometimes missed. When large groups of young goats are run together it can be difficult to notice oestrous behaviour. It is much easier when goats are being handled daily as with the milkers. If possible it is very useful to run a vasectomised teaser male with such groups of young goats. If a sheep raddle harness (Figure 5.8) is used on the male he will mark the oestrous females when he attempts to mount them. If the females are checked twice a day it will be possible to pick out those that are in oestrus and these can then be taken to the appropriate stud male.

As already discussed, a male running with females in this way may also be of benefit in inducing oestrus early and also in synchronising oestrus.

Those who experience most trouble in detecting oestrus seem to be those people with only one or two goats. This is possibly because in comparative isolation the typical oestrous behaviour and the interaction between goats at this time is suppressed. Those with large herds, particularly if there are male goats on the farm, rarely experience problems in this respect.

Mating

The strategy for mating goats will vary according to the interests of the goat farmer and the size of the herd. The simplest system is to let the male or males run with the females during the breeding season and after allowing 2 or 3 cycles, i.e. 6–9 weeks, it is assumed that all females that are likely to be mated will have been and the males can be removed. Kidding will be expected over a 6–9 week period 150 days from when the males were first put in with the females. If using this system 1 male will be required for every 30–40 females.

Such a system would be rarely used. Most goat farmers would wish, at least, to know when mating took place so that the precise kidding date can be predicted and also most people would want to put particular females to a particular male.

To be able to do all this it would be necessary to group the females according to the male that one wanted to use. These would be put together in a pen, paddock or field. To be able to time matings a sheep raddle harness as shown in Figure 5.8 could be used and thus the females would be marked as they were mated. If the goats are checked twice a day it would be possible to record the goats mated each day.

If the raddle crayon is changed to a different colour every 20 days it will be possible to detect those matings that were unsuccessful as these goats will be mated a second time 21 days after the first.

This mating system is exactly the same as used for most commercial sheep flocks and could be used with extensive goat systems such as those with large Angora or cashmere herds.

For pedigree mating, and certainly where a single female is brought to a particular stud male, a hand-mating technique will be used. This simply means the female will be led to the male who would normally be brought out of his pen onto a convenient level piece of ground nearby.

If the male is working well and the female is properly in oestrus mating will usually take place very quickly. However, a male may spend some time going through courtship behaviour which may involve much rubbing against and spluttering over the female. This behaviour should not be constrained in any way as this could jeopardise the chances of success.

When the male mounts the female a good sign of a successful mating is if he throws his head back as this is the normal sign of ejaculation having occurred. If the male spends a lot of time rubbing and spluttering without mounting the female it probably means the female is not in oestrus. If the male has behaved like this with a number of females his performance and ability must be suspect.

Figure 5.9 shows a simple chart that can be kept on the wall of the goat house for recording the events at mating. Such a chart could be used in any situation where timing of mating is possible. Space is provided for a number of matings as it will take two or three to show up the problems just described.

As can be seen in the example given the female Earley Princess did not mate successfully even when a different male was used and it is probable that she was infertile.

In the case of Earley Pat it would appear that the male Earley Lion was at fault as she mated successfully with Earley Mike. In fact it can be seen that Lion did not successfully mate any females and, therefore, it is possible that he was no longer fertile.

The chart also has spaces for the predicted kidding date and thus it would be possible to keep a check on the pregnant females as they come near to their due kidding time. The number of kids can also be recorded along with any other important information such as veterinary treatment. All of this information can be transcribed into the herd records which may be in a card index or on computer files.

Whichever system of mating is used goats are normally fertile animals and a conception rate at the first mating of 80–90 per cent can be expected during the natural season. Goats that have been induced into oestrus out of season using hormone treatment, such as progesterone sponges, usually have a lower conception rate.

Female	Male	1st mating	2nd mating	3rd mating	Date due to kid	Date kidded	Kids	Remarks
Earley May	Earley Mike	9/10/87			9/3/88	10/3/88	1♀ 1♂	
E. Mary	Mike	11/10/87			11/3/88	10/3/88	1♀ 2♂	1♂ born dead
E. Min	Mike	15/10/87	5/11/87		5/4/88	6/4/88	2♀	
E. Nice	Mike	7/10/87	28/10/87		28/3/88	28/3/88	3♀	1 kid small
E. Nancy	Nick	21/10/87			21/3/88	19/3/88	1♀	
E. Pat	~~Lion~~	3/10/87	24/10/87	15/11/87 MICK	15/4/88	17/4/88	2♂	
Earley Princess	~~Lion~~	4/10/87	NICK 25/10/87	NICK 17/11/87	17/4/88	—	—	not pregnant
E. Phil	~~Lion~~	20/10/87	NICK 10/11/87		10/4/88	13/4/88	2♀ 1♂	
E. Paula	Mike	3/10/87			3/3/88	4/3/88	2♀	

Figure 5.9   A breeding chart for the day-to-day recording of mating and kidding.

Artificial Insemination

Artificial insemination has been practised in goats for a long time.[4] However, it is only comparatively recently that the technique of freezing semen has been developed.[5] In Britain the company Caprine Ovine Breeding Services Ltd (COBS) was formed in 1981 to develop an artificial insemination service particularly for goats. The first few years of this company were spent developing collection, processing and insemination techniques and now semen can successfully be diluted and frozen in liquid nitrogen at $-196°C$ and experienced inseminators are achieving conception rates at first service of over 60 per cent.[6]

COBS has trained more than 80 people in goat AI techniques and although not all have bought the necessary and expensive equipment a nationwide service is beginning to develop.

AI has many advantages, many being the same as for cattle, and the technique could revolutionise the goat industry in the same way as it has for cattle. The advantages of AI for goats are:

1. Possible easier access to top quality males thus improving the overall quality of the national herd.

Figure 5.10   Collecting semen from a Bagot male.

2. Semen can be used even after the males are dead and certainly after a male's value has been assessed through the performance of his progeny.
3. AI reduces the risk of disease as it removes the need to move goats to and from farms holding stud males.
4. Male goats may be a problem to keep for small-scale goatkeepers. AI removes the need to do so.
5. Large groups of synchronised females can be inseminated at the same time.
6. The quality of UK dairy goats is renowned throughout the world and thus there is a good potential export market for semen from good males.

Of course there are disadvantages but these do not reduce the value of the technique and it is highly likely that as the interest in goat farming increases so will the use of AI. The disadvantages of AI for goats are:

1. Conception rates from AI would not be expected to be as good as with natural mating.
2. Special training and expensive equipment are required.

3. If a trained inseminator is not located nearby travelling costs can make the service expensive.

The most common method of restraint for AI involves holding the goat's back legs off the ground and presenting her rear to the inseminator. To do this the owner/handler stands astride the goat's neck, facing the goat's rear. The inseminator lifts the goat by her hocks and brings her up so that the handler can hold the hocks tight up against the goat's lower abdomen pulling her up against his/her chest. (Figure 5.11). It is best if the handler can lean into and rest his/her back into a corner. By doing so it is possible to restrain even quite large goats for insemination.

The inseminator uses an instrument called a speculum to look into the vagina of the goat to locate the entrance to the cervix (Figure 5.12). If the goat is at the correct stage of oestrus the cervix may be slightly dilated and it is sometimes easy to insert the AI 'gun' some distance into the neck of the cervix. However, the muscular bands, described earlier in this chapter, can make it difficult to insert the gun into the cervix at all in which case semen is splashed onto the entrance. The chances of conception will be reduced if this happens.

Those who have inseminated cattle say that goats are easier because the cervix, etc. can be seen, whereas with cattle it is all done by feel. Those who have never inseminated anything think goat AI is difficult!

For AI through the cervix, the semen is frozen in 0.5 ml plastic straws. The semen is diluted so that each straw contains approximately 120 million sperm. The amount of semen collected from a male varies but averages about 15–20 straws per ejaculate.

An insemination technique, using an instrument called a laproscope, deposits semen directly into the uterus through the body wall. This more positive siting of the semen dose allows less semen to be used to achieve conception rates at least as good as when the cervical technique is used. A disadvantage of this technique, in Britain, is that it can only be carried out by a qualified veterinarian.

Embryo Transfer

In recent years the techniques for preserving fertile embryos and transplanting them into suitable recipient animals that will become surrogate mothers have been considerably developed. Until recently much of the work on farm animals was concentrated on cattle. The technique is particularly attractive for rapidly increasing the number of progeny that can be produced by one female in a single breeding season.

It has been used most in breeds and species that, usually for reasons of scarcity, are valuable and, therefore, where the relatively high cost involved can be justified. In the case of goats embryo transfer has been used with Angoras and to a lesser extent with improved cashmere-producing crossbreds.

Embryo transfer involves treating a goat with a series of injections of Follicle Stimulating Hormone (FSH) and Luteinizing Hormone (LH) to synchronise oestrus and to induce super-ovulation. The goat is mated repeatedly throughout the peak of her oestrus period in order to increase the chances of a large number of the ovum, 12 on average, being fertilised.

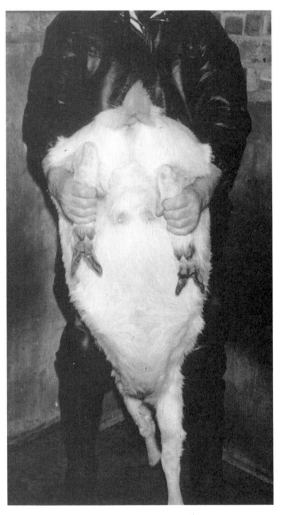

Figure 5.11   Presenting a goat for artificial insemination
(courtesy Food Research Institute, Reading).

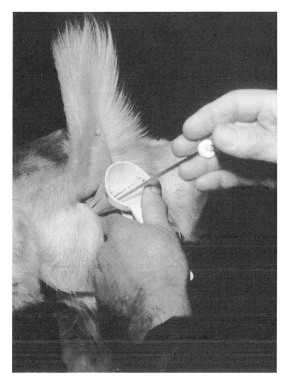

Figure 5.12 Artificial insemination using a speculum
(courtesy Food Research Institute, Reading)

At the same time a number of ordinary goats, usually 7–10, are also treated with hormones so that their oestrous cycle will be exactly synchronised with that of the donor.

Six days after the donor is mated she undergoes a small operation to exteriorise the uterus which is then flushed to recover any fertile embryos that may be present. The embryos that are collected are examined under a microscope by an expert embryologist who can determine their quality and, therefore, those that are fit to be used. Two of these will then be transferred by a similar operation to each of the recipient goats. Embryos can be frozen and stored in liquid nitrogen in a similar way to semen and thus they can be shipped around the world and can be used a long time after they were collected.

When considering the economics of embryo transfer it is important to remember that 12 embryos recovered will not mean 12 live kids born. There are many points in the programme where the embryos may die. Good synchronisation of the donor's and recipient's oestrous cycles is most important and stress and nutrition can effect embryo survival after implantation.

For every 12 embryos flushed 2 would probably be infertile or unsuitable for transfer for other reasons. Not all the recipients will be suitable for use and of the 5 that are, 3–4 will become pregnant. Taking all those variables into account the average number of kids born per flush will be 5–6 and of these it is normal to expect 50 per cent male and 50 per cent female.

Pregnancy Diagnosis

For economic reasons it is often useful to determine whether or not a goat is pregnant. Such a diagnosis can save food as a non-pregnant goat will be fed less and also if non-pregnant there may be time to try another mating or AI again before the end of the season. It may also be necessary if one is buying or selling goats that are supposedly pregnant.

Earlier in this chapter the hormonal control of oestrus and pregnancy was described and it was explained how the hormone progesterone, produced by the ovary, was responsible for the maintenance of pregnancy. It follows, therefore, that if the level of progesterone secreted can be measured at the time the goat would be expected to come into oestrus again and it was found to be high there is a good chance that she would be pregnant.

A simple test kit is available for testing progesterone levels in milk, and thus diagnosing pregnancy, in cows. Unfortunately, these kits do not work reliably for goats probably because there are substances in goat milk that interfere with the test. Even in cattle high levels of progesterone do not guarantee pregnancy as faults in ovarian function, such as caused by a cyst, can cause high levels in non-pregnant animals. False pregnancies will also give high progesterone levels.

Another milk test for pregnancy, developed for cows, involves testing for a metabolite produced by the placenta of pregnant animals called oestrone sulphate. As it is only produced by the placenta false positive results do not occur. The Milk Marketing Board Laboratories in Worcester offer a testing service and in the case of goats can test from a milk sample after 50 days gestation.

A recent development, particularly for sheep farmers, has been the establishment of businesses offering pregnancy diagnosis by ultra-sonic scanning. Some of these, particularly those using a rectal probe, are able to make a diagnosis after about 35 days and if left a little longer can usually count the number of foetuses. Some of these scans can also translate the signal into a visual image which, when photographed, can give a very useful permanent record of the diagnosis. This system seems to work equally well if not better with goats.

If a goat thought to be pregnant is visiting a veterinary surgery it may be appropriate to diagnose pregnancy by X-ray. Foetal bones show up after about 85–90 days and of course with a good X-ray the number of foetuses should be clearly visible.

During the last 6 weeks of gestation, i.e. after 110 days it is often possible to see the foetuses moving particularly when the goats are lying down. Also at this time the foetuses can sometimes be felt by firm palpation deep into the lower abdomen just in front of the udder.

Pregnancy (Gestation)

Pregnancy or gestation lasts for 150 days in the goat but a few days variation either way is quite normal. Once one is certain that goats are pregnant, or if in a large herd it is assumed that the majority in a group are pregnant, a change in management is appropriate in order to minimise kid losses. Care should be taken to ensure that there is no excessive bullying among pregnant females particularly when horned and polled goats are mixed.

The level of feed given to pregnant goats is important and should be increased, as described in Chapter 4, during the last two months of gestation. Feeding during this period will influence the size of the kids, the mohair yield of Angora kids, the development of the udder and subsequent milk yield and the female's forage intake during lactation can be conditioned at this time.

During the last few weeks of gestation the udder will undergo rapid development and in the case of heavy milkers may look very swollen and engorged. There is often a temptation to milk out a little milk at this time to relieve the pressure but unless the goat is in obvious discomfort it is better not to do this as it will affect the production of colostrum at kidding.

The development of the udder is influenced by a number of hormones including progesterone from the ovaries, prolactin from the pituitary gland and placental lactogen from the placenta. This latter hormone, not surprisingly, is produced in greater quantities according to the amount of placental tissue. Thus goats carrying a number of foetuses are likely to produce more milk than those with single foetuses.

Kidding

A kidding first-aid kit should be assembled ready for the kidding season. This should include a clean bucket, mild soap, obstetric lubricant, paper towels, broad spectrum antibiotic, antiseptic and a 'lamb-reviver'/stomach tube.

It is a good husbandry practice to put goats close to kidding into a separate clean pen, but still within sight and sound of others. These kidding pens can be of a temporary nature, at their simplest constructed from some hurdles in the corner of the main pen. There should be no buckets or other projections in the pen that could injure the goat when she is straining to give birth.

In reasonable weather goats will kid outdoors but indoor kidding will almost certainly result in less kid losses. Similarly individual kidding pens need not be used but if they are it is possible to operate a higher standard of husbandry and again losses will be reduced. Unlike sheep, goats do not seem to have problems with kids becoming mixed and often adopted by the wrong mother in communal kidding pens.

It is usually possible to judge that kidding is imminent a few days before it actually happens. The goat will spend more time lying down and may be observed not to be cudding as often as may be expected and may look generally rather uncomfortable. The udder, particularly of heavy milkers, will look very full.

Usually within 24–48 hours of kidding the pelvic bones relax to facilitate the passage of the kid. This relaxation of the pelvis is often termed the breakdown of the tailhead because hollows appear either side of the base of the tail. This can happen much earlier in some goats or may not be seen at all in others.

Within 24 hours before kidding the goat will tend to draw away from her companions and will seek a quiet corner of the pen. Here she will paw at the ground and will repeatedly lie down and stand up again. A discharge or even the foetal membranes containing the amniotic fluid may be seen at this time.

Once a goat has started to strain a regular check should be made to ensure there are no problems. If she has been straining for an hour without any sign of progress it should be assumed that there are problems. This will most likely be due to the incorrect position and presentation of the kid.

With a normal presentation the kid will be born head first with its front legs stretched out in front. The first things to be seen, therefore, will be the hooves of both front feet and the nose (Figure 5.13). If only one foot, no feet or no head can be seen it usually means they are bent back and the kid will have to be pushed back to enable the bent limbs or head to be pulled forward.

People with no experience of these emergencies are advised not to attempt to help. The kidding season is a time when it is good policy to cultivate a friendship with an experienced local goat-keeper or with a local sheep farmer. Well-meaning, inexperienced people can do a lot of damage trying to assist difficult births and it is far better to be able to call on the

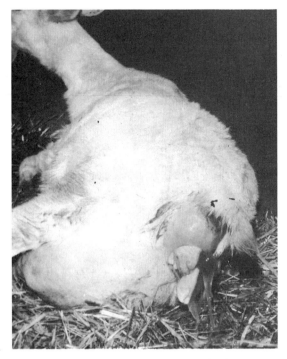

Figure 5.13 Kidding, a normal presentation (courtesy Food Research Institute, Reading).

help of someone with experience who can then be watched to gain confidence for future emergencies.

As goats are usually prolific, with dairy breeds in particular often averaging more than two kids, there is more chance of kids becoming entangled during a difficult birth. In this situation traction can be mistakenly applied to a right and left front leg only to discover that they belong to different kids. This sort of thing is dangerous and could result in the death of both mother and kids.

As soon as the kids have been born, and remember there could be three or four and quins have been recorded on several occasions, it should be ensured that the mother licks them clean. She is most efficient at doing so and this activity is important for developing the mother-kid bond. If membranes become stuck over the kid's mouth and nose these should be removed as they could suffocate it.

It is most important that kids feed within 2–4 hours of birth. If they do not they will become too weak and will never feed and will die. If they become chilled this will also weaken them. The milk produced during the first few days after birth is called colostrum and is rich in fat, protein and antibodies. It is a wise precaution to keep some colostrum for emergencies in a deep-freezer. This can be warmed and fed to those kids whose mothers are unable to feed them.

Weak kids without the strength to suck their mother's teat should be stomach tubed using a 'lamb-reviver'. The small bottle is filled with warm colostrum which is simply gravity fed via the tube into the kid's stomach. A great many lives have been saved in this way and kids without the strength to lift their heads off the ground may gain enough strength to get up and feed from their mothers.

Immediately after birth it is a good practice to dip the kids' navels in antiseptic, usually iodine-based, to prevent infection. Infections that get in via the navel of new born animals often end up in the joints, giving rise to the condition known as joint-ill, and are extremely difficult to treat once there.

Once the mother and kids have been together in the kidding pen for a few days and it is obvious that all are doing well, decisions have to be made about kid management and this will depend on the nature of the herd and the products produced from it.

REFERENCES

1. Ricordeau, G. (1981), 'Genetics: breeding plans', in *Goat Production*, ed. Gall., C. (Academic Press, London).
2. Ashbrook, P. F. (1982), 'Year-around breeding for uniform milk production', in the *Proceedings of the 3rd International Conference* on *Goat Production and Disease* (Dairy Goat Journal Publishing Co., *Scottsdale, Arizona*).
3. Jarvis, Marjorie, B. (1987), personal communication.
4. Asdell, S. A. (1948), *British Goat Society Year Book* (British Goat Society, Bovey Tracey).
5. Mowlem, A. (1983). 'The development of goat artificial insemination in the United Kingdom', *British Goat Society Year Book* (British Goat Society, Bovey Tracey).
6. Clabburn, M. and Smith, J. S. (1988), 'Artificial insemination in dairy and Angora goats', *Monthly Journal of British Goat Society*, vol. 81.

Chapter 6

Kid Rearing

Once kids have had time to be with their mothers for 2–3 days decisions have to be made about what to do with them next. Usually dairy-goat kids will be removed and reared away from their mothers so that the milk can be sold or processed and fibre-producing breeds will probably be reared by their mothers. However, various factors will influence the decision and all possibilities will be discussed in this chapter.

If dairy goats are being farmed it is assumed that milk is the principal product even though it may be sold as liquid or as products such as cheese. Although there is no set pricing structure for goat milk, considering the average price being obtained by producers, there is a tremendous economic advantage in rearing kids on a milk replacer rather than goat milk.

Having decided that it is advantageous to rear kids artificially it is then necessary to consider why they are being reared. The reason may influence the system used. For example, kids reared for sale as potential breeding stock will sell at a premium if they are better grown than their peers in the sale ring. Kids reared for meat, when short-term costs will be important, may be reared differently from kids reared as herd replacements where the investment in rearing cost will be spread over a much longer period.

PENS/ENVIRONMENT

Whatever the reason for rearing kids the important criteria will be low mortality, good growth and reasonable economy. To achieve the first of these objectives every effort should be made to create a good hygienic rearing environment.

The environment, pen construction and heat sources necessary for rearing kids are described in Chapter 3. It is best if the kids are started off in small groups of not more than 12 and, therefore, they will need a pen of about 3.6 m². Groups of kids of similar size should be put together; otherwise bigger kids tend to dominate the feeders.

It is most important that the pens and, in fact, the whole kid-rearing area are kept very clean with plenty of good clean straw bedding in the pens. The pens should be cleaned out at least once a week and all feeders and water bowls should be cleaned daily.

MILK FEEDERS

There is quite a range of milk feeders, now called generically, 'lamb-bars'. It is important that they should be easy to clean thoroughly and quickly. Polythene and stainless steel are best from this point of view although stainless is likely to be prohibitively costly. Lids are important because kids spend a lot of their time up on their hind legs looking over the sides of their pens and in doing so would almost certainly deposit muck into open milk feeders. It is also important that the feeder can be fixed firmly to the sides of the pens or otherwise the kids will knock them off.

Buckets/Troughs

Kids can be fed quite successfully by bucket but this is only suitable if they are being fed a limited quantity of milk at each feed and if they are supervised when they are feeding. The system used for calves, where the buckets are fixed to the outside of the pen gate and the calves put their heads through an opening in the gate, will not work for goats. They will tend to escape through the head hole in the gate and will also foul the milk by resting dirty feet on the gates or buckets.

Buckets or troughs do have the advantage of being cheap and easy to obtain and some farmers will devise systems using them.

Feeders with Teats

If milk is to be offered *ad lib* a closed feeder is required so that the milk, which may be enough to last 24 hours, will not become fouled or contaminated. There are two basic types of teat-feeder units that will fulfil this function. One type has the teats fixed directly into the milk container and with the other a small tube is attached to the back of the teat so that milk is sucked up from the bottom of the container. Small kids seem to have difficulty in sucking the milk up the tubes of the second type even when ball valves are fitted to stop the milk flowing back down the tube.

The disadvantage of the teats being fixed directly into the milk container is that bigger kids may pull them out and then all the milk will run out. However, if this becomes a problem it is usually an indication that the kids are big enough to be weaned. A useful teat for this type of feeder is made by Ripper Systems (Bedford). This teat has a large flange which, if

Figure 6.1   A gravity milk feeder

the correct-sized hole is used, makes the teat very difficult to pull out of the feeder.

Automatic Milk Feeding Machines

For rearing kids on a large scale some farms now use one of the automatic calf-feeding machines that are now available (Figure 6.3).

These machines mix and supply warm milk on demand when kids suck teats mounted on the side of their pens and which are connected to the machine by narrow plastic tubes. These machines allow substantial savings in the labour required for rearing kids.

One problem that has been found is that, because the machines dispense warm milk, there is a risk that the kids may drink too much. If this happens they may suffer diarrhoea or scouring. This causes dehydration and this then causes them to drink more and one then ends up with a very difficult situation to get out of. Various things may trigger off this problem

Figure 6.2   A milk feeder with the teats attached to plastic tubes.

such as a sudden fright after which the kids may run to the teats and drink in the same way as they would run to their mothers for comfort.

Whichever type of feeder is used it cannot be emphasised too much how important cleanliness and good hygiene are when rearing kids and all feeders should be washed out daily. In the case of the automatic milk-feeding machines the pipes and teats should be regularly cleaned in accordance with the suppliers' instructions.

TEACHING KIDS TO DRINK

Although not so much of a problem as lambs or calves the longer kids are left with their mother, the longer it will take for them to learn to feed from a milk feeder. It may be that, with just a few kids, they could be intro-duced to a teat by way of a bottle as a step towards the artificial feeder. At

Figure 6.3   An automatic milk feeding machine (courtesy of British Denkavit Ltd).

the very least it will take about 24 hours and a great deal of patience before one can be sure that a kid is feeding well. It is much easier to get kids to feed from a feeder if others in the pen have already learnt to feed. All the kids will tend to crowd round one another and if one or two are drinking the rest will soon get the idea. Even if a milk replacer that is formulated to be fed cold or at ambient temperature is used, it is advisable to warm it slightly when teaching young kids to drink.

If any kids are much smaller than the rest these are best put into a separate pen where they can be given special attention and where they will not be bullied or prevented from feeding by the bigger kids.

CHOICE OF MILK REPLACER

Several feed companies manufacture a range of milk replacers of different composition which means a goat farmer is faced with a bewildering choice. If we look at the composition of goats' milk it will be seen to be very similar to cows' milk in terms of its gross constituents such as fat, protein and solids and it has virtually the same energy value, whereas sheep's milk has a much higher fat and solid content. From this it would be logical to feed a milk replacer formulated to replace cows' milk, in other words one for calves.

A great deal of work was done at the former National Institute for Research in Dairying comparing feeds and feeding systems.[1] It was concluded that the earlier kids were weaned the less important was the milk feed. Small differences in performance, observed during the few months the kids were fed milk replacer, were not at all apparent after the kids were weaned and had been on solid food for a few months. Even where there was a significant difference as, for example, between a group on lamb replacer and one on soya and whey replacer it was considered that the increase in cost outweighed the advantage of faster growth in the better group.

Some manufacturers formulate milk replacers specifically for goat kids as a result of demands for a product without the growth promoters normally included for calves. The limited demand for such a special product inevitably means it will be more expensive than a calf replacer that is made in hundreds of tonnes at a time. It is unlikely, therefore, that these special products have any relevance to goat farming because of the vast increase in rearing cost. At the time of writing milk replacers vary in price from about £700 to over £1,000 per tonne.

Whichever milk replacer or milk feed is used it is most important not to change the feed as this will almost certainly cause the kids to scour. They may soon get over this but it can so often be the start of a major health problem and should, therefore, be avoided if at all possible.

WATER AND SOLID FEED

Kids will start to pick at solid food such as hay and straw from about 10–14 days of age. If they are to be weaned early they should be encouraged to eat solids and should be given a little concentrate feed in addition to good quality hay each day. They will only eat a little at first but by the time they are weaned they may be eating 400–500 g/d.

Any feed formulated for calf or lamb rearing is suitable for kid rearing and if it is policy to feed the adult goats a pelleted or cubed concentrate then pellets or cubes should be fed to the kids.

It is always difficult developing or finding a method for feeding kids concentrate in a way where they cannot get in it and foul it. Some form of trough with a hood partially covering it may be reasonably successful but, as with most things for kids, if they can get their heads in they can usually get the rest of their bodies in as well!

Hay should be offered in some form of rack that is off the floor. Hay nets should not be used as the kids will use them to play 'king of the castle' and legs may get caught in the netting and may get broken.

REARING REGIME

The choice these days is between feeding a ration of milk replacer over a relatively long period, 10–12 weeks, or weaning early at 6–8 weeks and using an *ad lib* milk replacer. In Table 6.1 the two systems are shown with figures in brackets showing alternative ages.

If kids are to be weaned as early as 6 weeks care must be taken to ensure they are eating solid food adequately and are drinking water. Well-grown kids on a regime that gradually reduces the amount of milk offered, during the 2 weeks before weaning, should suffer minimal or no post-weaning check in growth rate.

Some farmers replace the milk with water as weaning approaches as a way of getting the kids to drink. This may be a useful practice, particularly with the machines, but there is also a danger that the kids will be difficult to teach to drink from something other than a teat later on.

There is a significant difference in the feed costs of the two systems. The importance of this would be related to the reason for rearing the kids. If they were being reared for meat then cost would be critical. If they were being reared for herd replacement then a good steady growth may be more

Table 6.1 Alternative artificial kid rearing regimes

Early weaning at 6 or 8 weeks*		10-week weaning	
Age (weeks)	Number of feeds per day	Age (weeks)	Feeds per day (ml)
4 (6)	ad lib	1–6	3 × 750
5 (7)	half amount consumed last day of week 4 (6)	7–8	2 × 850
6 (8)	half amount consumed last day of week 5 (7)	9	2 × 570
7 (9)	no milk	10	1 × 570
		11	no milk

* Figures in brackets for 8-week weaning

important than the cost which, in terms of investment in the goat, will be spread over a much longer period.

Similarly if kids are being reared for sale, rearing costs may be a small fraction of the potential value of the kid. Anything that makes a kid look better than others in a sale is well worth the extra rearing costs. This is particularly true of the currently high-priced Angoras.

In general terms early weaning will cost less in terms of feed. Feeding milk replacer *ad lib* will give the fastest growth rate. Kids that are reared artificially using an *ad-lib* feeding system will grow faster than those that are naturally reared by their mothers and the cost of this faster growth rate will only be justified if well-grown kids, for whatever reason, have a high value.

POST WEANING

Every effort will have been made to ensure that the kids are eating their solid food and are drinking satisfactorily. It is worth going to some trouble to achieve this if a post-weaning growth check is to be avoided. Fresh concentrate feed should always be given each day even if this does mean some wastage (there are usually other animals on the farm to eat this). Hay should be of the best quality with no mustiness or other contaminants that would put a kid off eating it. Similarly water should be kept clean and should be provided via drinkers or troughs that are easily accessible. If it is difficult to adjust the height of drinking troughs, a couple of concrete blocks can be provided to give the kids a step up.

It is a good practice to change the kids over to the concentrate feed used for the adults at around 6 months of age. By this time they will be well grown and able to cope with checks in growth, caused by a drop in consumption, until they get used to the new food. Also, if these changes are made much later, poor food consumption may affect a subsequent pregnancy or lactation. Kids are usually less of a problem when introducing a new feed than adults.

Once the kids have been weaned and are feeding and growing satisfactorily they can, if necessary, be moved out of the 'nursery accommodation' and into the more open building as used for the adults. There is still not much justification for putting the artificially reared kids out to grass. They will be a problem to keep fenced in. They will spend much time playing and exploring their new environment rather than eating and growing and they may well suffer from heavy infestations of internal parasites. All this would mean, to some extent at least, that the advantage gained by the artificial rearing regime and consequent good performance could be lost if the kids were then turned out to grass too soon.

If the kids are born in the spring they would not be turned out, if at all, until they are about one year old and when the rest of the herd are turned out after spending the winter indoors.

This standard of husbandry is appropriate and economic for dairy goats and perhaps for fibre-goat-breeding stock but would not be economic for commercial-fibre-producing herds.

NATURAL REARING

Artificial rearing would not normally be considered economic for commercial-fibre-producing herds. In such herds it is likely that it will be confined to orphan kids or those for whatever reason that cannot be reared by their mothers.

Once they have been removed from the kidding pens the kids and their mothers would be put together in suckling groups. These may be within a building or covered yard if the state of the weather and the pasture does not allow them to be turned out. Wherever they are it is advisable to devise a creep system that will allow kids to move into another pen independently of their mothers, to give them access to a little extra food.

The creep can be simply a hurdle with gaps big enough to allow the kids, but not their mothers, through. It is much more difficult to design creeps for goats than it is for sheep. Adult goats seem extraordinarily adept at getting through gaps that are seemingly only big enough for kids to pass through.

This same principle can be used when the goats are out to grass. The creep will not only allow the kids to be fed a little extra concentrate but they will also have access to fresh grazing.

It is likely that kids out to grass will face a heavy challenge from internal parasites, particularly stomach worms, and, therefore, regular drenching every 3–4 weeks is advised.

Naturally-reared kids would normally be weaned by removing them from their mothers at about 5–6 months of age which, with natural kidding patterns, would be just before the breeding season started again. By this time the kids would be managed more or less like adults. It should be remembered that entire males may need to be separated from females at 3–4 months of age, particularly if this coincides with the beginning of the breeding season. The economics of this and the other rearing systems are discussed in Chapter 13.

MARKING FOR IDENTIFICATION

In any well-run herd individual goats will need to be positively identified so that a record of their performance can be kept. It will also be necessary to record their relationship with others in the herd if an improvement in the performance of the herd is to be achieved by selective breeding. To achieve this it will be necessary to mark kids for identification before they leave their mothers and before they are mixed with others.

The only truly permanent method is to tattoo the kids, usually in the ear. Ear tattooing is a pre-requisite of registering goats with the two main breed societies, the British Goat Society and the British Angora Goat Society.

Some people are reluctant to earmark small kids that are removed from their mothers at only a few days old. Although it is quite satisfactory to earmark at this age it is also possible to use a simple colour code, applied using an aerosol marker to the kid's back. There are many simple examples of such a code, one of which is shown in Figure 6.4.

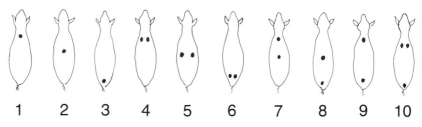

Figure 6.4   A simple spot marking code for kids.

A number of tattooing tools are available from most animal or veterinary equipment stockists. (Figure 6.5.) Most work on the principle of small numbers made up by small sharp needles that are fixed into a metal base. Each number can be slid into position in the jaws of the tattooing tool. It is sensible to test each number combination by 'tattooing' a piece of thin card before earmarking the kid.

Once the number has been indented in the ear special tattooing paste is rubbed into the wound using a fine toothbrush or a piece of cotton wool. Some people like to apply the paste first and although effective this does make a mess of the needles and it is important for the sake of hygiene to keep these clean. The tattooing needles should always be washed and disinfected between use and great care must be taken when a tattooing tool is shared between goat-owners or between the members of a goat club.

Figure 6.6 shows the tattooing system used by the BAGS – the BGS system is very similar. Both use a letter to denote the year of birth and both use a letter coding for each breeder's herd.

Figure 6.5   A tattooing tool.

Figure 6.6   An ear tattoo as
required by the British Angora Goat
Society. B = grade; RH = herd;
12 = number of goat; Z = year born.
The British Goat Society's system is
the same but would not have a letter
indicating the grade.

The Angora Society also has a register for commercial goats that will not be registered as pedigree breeders. If such goats are sold, the number 888 is tattooed into the ear to obliterate any number already there and prevent a commercial grade animal being disguised as one that is registered as pedigree.

For quick field identification coloured plastic ear tags can be used. These can be embossed with much the same information as would be in the tattoo but, in addition, the many colours available can be utilised for a simple code to aid management such as year of birth or bought in goats. Unfortunately goats have a bad record when it comes to man-made adornments and one would expect around 25 per cent of tags to be pulled out within the first 2 years. It is also difficult to tag goats that have complicated ear tattoos without obliterating the tattoo.

A variety of other identification aids are available and although effective and useful for other stock few are of any use for goats. If there is any way that a goat can remove a tag, collar, leg band, etc. it will and being very active animals there is always the danger that some of these devices, such as collars and bands, may get caught up with obvious serious consequences.

Coloured goats can be freeze branded and if on the rump this is useful for identification in the milking parlour. White goats can have a number painted on using an aerosol marker spray and, although this will not last more than a few weeks, it will be useful in the milking parlour.

CASTRATION

Although it is becoming accepted practice to rear entire male animals for meat these days, there is a particular problem with male goats in that they are so smelly during the breeding season. Also young male goats are very sexually precocious with fertile matings being recorded at 3 months of age. For ease of management it may, therefore, be desirable to castrate them if they are to be kept beyond this time.

If males are definitely not to be used for breeding it is normal to castrate them using the rubber ring method. This must be done within one week of birth. The method involves the use of a special tool for stretching a small thick rubber ring which is then placed over the scrotum and it is left in place at the base, making sure both testes are in the scrotum before the ring is released from the applicator. The scrotum will shrink and die and will drop off after a few weeks.

If it is possible that the kids may be required for breeding it will be necessary to leave them entire until their quality as potential stud males can be assessed. In the case of Angoras this will be when they are 18 months old. At this age it would be normal to castrate surgically which involves slitting open the scrotum and removing the testes. This must be done under anaesthesia and many will no doubt prefer their vet to carry out this operation.

Another method involves the use of a Burdizzo or bloodless castrator. This is an instrument rather like a large pair of carpenter's pincers with blunt jaws. These jaws are used to crush the blood vessels and spermatic cord at the base of the scrotum. If effective the testes should feel small and hard a few weeks after the operation.

DISBUDDING

In most intensive dairy herds these days the goats are disbudded. Horns have many disadvantages and few advantages. They often result in heads becoming stuck in fences, etc. and quite a lot of damage may be done through fighting. Horned heads are often a problem in feeding yokes such as those used in most milking parlours. At present Angora and cashmere goats are not disbudded and it is probably not necessary when they are kept extensively outside.

Since 1982 it has been illegal for goat-keepers or other lay persons in Britain to disbud goats. It can only be done by a qualified veterinarian. It is, therefore, inappropriate here to go into detail about the procedure and veterinarians are recommended to read the article on the subject by Buttle et al. in the journal *In Practice*.[2]

There are two problems with goat disbudding. First of all the nerves to the horn bud are more difficult to block than in calves because the anaesthetic must be injected in two precise sites either side of the orbit. Also the horns of goats grow much more vigorously than those of calves and, therefore, a different technique is required using a much hotter disbudding iron. Female kids are a little easier to disbud than males, in fact it is quite unusual to see a mature male that has not got, at least, a small piece of regrown horn or scur. Badly regrown horns are usually very misshapen and often curl over and will dig into the goat's head. If this is the case they will need cutting off with a hack-saw every now and again. Care should be taken to establish which parts of such horn still carry blood vessels as profuse bleeding can result from a horn trimming operation.

The dehorning of adult goats is not recommended. It is very traumatic for the goats – they usually bleed a lot and often the sinuses behind the eyes are opened up and there is a real danger of the wound becoming infected. If it really must be done it should be carried out during the winter when there are no flies about.

REFERENCES

1. Mowlem, A (1984), 'Artificial rearing of kids', *Goat Veterinary Society Journal*, vol. 5, no. 2.
2. Buttle, H., Mowlem, A. and Mews, A. (1986), 'Disbudding and dehorning of goats', *In Practice*, vol. 81, no. 2.

Chapter 7

Health and Disease

It is not appropriate in this book to go into great detail about the many health problems that could affect goats. That would be more appropriate for a veterinary textbook. However, there are many aspects of routine care that must be the concern of the goat farmer and there are a few important disease conditions that must also be considered.

When in comparable environments goats are probably healthier than other farmed species and do not suffer from any particular diseases that could be described as a major problem. All farmers are advised to develop a health-care programme for their goats in conjunction with their veterinary practitioner. Although this will involve some expense the savings made in having a healthy herd with minimal losses and a good level of production will make such a programme a very good investment. All vets with goat-farming clients should be persuaded to join the Goat Veterinary Society (see Appendix 2).

FOOT TRIMMING

As with most domesticated ungulates (cloven hoofed) a goat's feet grow continuously and, unless on particularly hard ground, need regular trimming. The technique of cutting away the excess hoof is exactly the same as for sheep and the main points are illustrated in Figure 7.1. The method for restraining most goats is different from sheep, however. Goats are not cast onto their backs like sheep, with the exception of Angoras, but are restrained in a standing position and the leg is held in a similar position to that of a horse being shod by a farrier. Angoras are broader and have a thick coat which enables them to be cast and held between the handler's legs in the same way as sheep. Even they have to be inclined to one side so that their weight is resting on the upper thigh muscles rather than on their backbone.

To keep feet in good shape when goats are on relatively soft or wet ground it may be necessary to trim their hooves every 2 months. This may not be practical on a large farm in which case goats that are obviously lame should be attended to and an effort should be made to have the feet

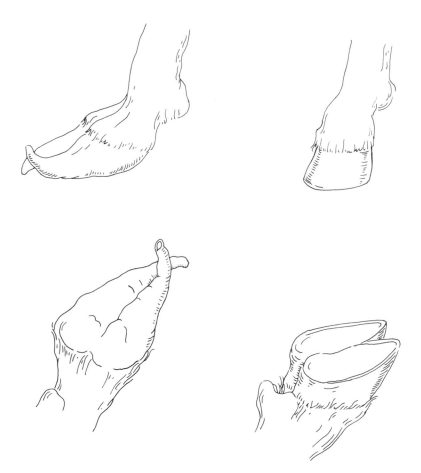

Figure 7.1   A badly overgrown hoof on the left, and, on the right, how it should look after trimming

in good shape when the goats are in the last half of pregnancy. Long hooves will tend to make a goat lame and are much more likely to become infected with the foot rot bacteria. This thrives in hooves that have wet mud trapped under overgrown nails.

If rot is a problem, which would be unusual in goats, they can be run through a foot bath containing a 4 per cent formalin solution. The foot bath must be cleverly designed because some goats will manage to walk down it without touching the liquid by balancing on the sides of the trough! This can be overcome by putting a low 'limbo' bar over the bath which will force the goats to duck and they will have to put their feet into the formalin solution. It is important to use the foot bath after foot trimming because it makes the hooves very hard.

Figure 7.2 Foot trimming (courtesy of the Food Research Institute, Reading).

INTERNAL PARASITES

These are a potential problem to most goat herds, particularly those with goats continuously grazing a small field or paddock. Goats can be infested with all of the common intestinal and stomach worms such as *Ostertagia*, *Haemonchus* and *Trichostrongylus*. Tapeworms, *Eimeria* species, may also infest goats but fluke and lungworms seem to be less of a problem.

Most of these parasites have a life cycle that involves mature worms reproducing inside the goat and producing eggs which pass out in the dung. These develop into many larvae which live off micro-organisms. These larvae then change to a dormant stage, sometimes a cyst, until they are eaten by the host animal, a goat. The larvae can survive on the ground for long periods including over winter. Weather conditions, to some extent, affect the survival of worm larvae; warm damp conditions are best and hot dry conditions least favourable.

The examination of faeces for worm eggs, by a veterinary laboratory, is a very positive method of diagnosis of heavy worm infestations. Goats with heavy worm infestations will look sick. They will look dull, listless, possibly anaemic and will probably have a watery diarrhoea (scour). It is best to assume worms whenever adult goats scour as this is always the commonest cause. Faecal egg counts will tell if the diagnosis is correct.

There is some evidence that suggests goats do not build up the same degree of resistance to worms as sheep. Therefore, although they may not be showing clinical signs of a heavy worm burden, they may be infested to a level that would be affecting their performance in terms of growth or milk production. It has been shown that milk yields could be increased by more than 17 per cent, over a 3-week period, after effective treatment against worms.[1]

Treatment of goats to prevent heavy worm infestations is partly a question of goat and pasture management with the appropriate use of modern anthelmintic drugs. Most anthelmintic drugs have a withdrawal period, usually up to 3 days after treatment, during which milk or the flesh from the goat cannot be consumed. This factor has resulted in many of the larger dairy goat units zero grazing or permanently housing their goats as this will reduce the risk of parasite infestation. It is necessary, therefore, to consider the two management strategies separately.

If goats are to be permanently housed or yarded they should be drenched 8 hours before being moved into clean premises and then again 14 days later to catch any immature larvae which might have survived the first treatment. Once established on this type of regime the risk of high worm burden should be minimal. However, infestations may still occur and therefore it may be sensible to drench zero-grazed goats at least at parturition (kidding) and possibly once more at peak lactation.

For goats that turned out to graze during the spring-autumn period a different treatment strategy is required. The peaks of worm burdens usually occur in spring, mid-summer and early autumn. The spring rise is due to the goats' ingesting overwintered larvae plus the rise in faecal egg count immediately after kidding. The goats should be treated when being turned out in the spring and then again during mid-summer when they must also be moved to clean pasture.

These days clean pasture, which means pasture that has not had sheep or goats grazing on it for 12 months, is hard to come by except perhaps on large mixed farms where crops can be rotated with grazing. Without such clean grazing it would be necessary to drench goats every 3–4 weeks to achieve effective control.[2]

Kids are particularly susceptible to internal parasites and their growth will be inhibited if good worm control is not achieved. In Chapter 6 the strategy for the indoor rearing of kids, because of this problem, was discussed.

There is a wide range of anthelmintic drugs on the market with most of the veterinary pharmaceutical companies making a variety of suitable products. Most will be produced for sheep and if goats are mentioned at all on the instruction leaflet, it will be suggested that they should be dosed at the same rate as sheep according to body weight. There is a little evidence that suggests higher doses of some drugs may be required for effective treatment of goats. Unfortunately goats are not yet of great economic importance for the necessary experimental work to be carried out to establish the true facts.

Special drenching guns are available from most suppliers and these are designed to make the procedure trouble free and these guns can be adjusted to give a predetermined dose. For small numbers of goats the drench can be satisfactorily administered via a small narrow-necked bottle such as a 'Coke' bottle.

Another parasite that may cause problems, particularly in kids, is Coccidia. This is a single-celled (protozoan) parasite which may invade the intestine or the liver. Like most internal parasites it has a complicated life cycle and infective oocysts are passed out in the faeces, a useful diagnostic feature. All adult goats carry this organism and it normally is not a problem. However, sick and young goats will be more susceptible and in these the Coccidia organisms may proliferate and cause the disease coccidiosis.[3]

It is perhaps commonest in kids reared in moderately intensive conditions at about 6–10 weeks of age and if kids scour at this age it may be best to start treatment before a positive diagnosis has been made. It can be treated by administering the drug amprolium ('Amprol', Merck, Sharpe and Dohme Ltd), in the drinking water. Drugs that can be administered in the feed are also available.

External Parasites

To the relief of fibre-goat farmers, goats rarely suffer from sheep scab and do not have to be compulsorily dipped. They may get a number of other ectoparasite problems that, while not so serious, will need remedial action and for some, preventative measures are possible.

All goats may be infested with lice and if the infestation is heavy the goat will lose condition and, in the case of fibre breeds, the coat will drop out at least in patches if not over the entire body. Normally lice are difficult to see even when numerous and, therefore, it is suggested that the goats are inspected regularly to prevent heavy infestations.

Most parasites seem to prefer those parts of the body that are fairly inaccessible such as behind the ears and on the inside of the upper limbs. With goats that are handled regularly the lice may be noticed, especially on white goats. The lice will be seen as minute cigar-shaped creatures moving about at the base of the hair. It is more difficult to see them on coloured goats and here it is more likely that the eggs (nits) will be seen as clusters of little white specks. A microscopic examination will easily identify both lice and their eggs.

The other serious ectoparasites are mites which cause a condition called mange. It is a mite that causes the sheep-scab problem mentioned at the beginning of this chapter. Most mites burrow deeply into the skin and cause a variety of different lesions depending on the species concerned. The chorioptic mite causes a diffuse reddening of the skin which supporates (produces fluid) causing extensive scabs. These tend to appear first around the feet and are spread to the head and neck region when the goat scratches itself. Chorioptic mites, causing chorioptic mange, often burrow very deeply into the skin and are, therefore, difficult to find in skin scrapings taken for diagnostic purposes.

Demodectic mange appears as localised pustular lesions which are caused by the mites invading the hair follicles and the sebaceous glands in the skin.

The other most serious mange in goats is caused by sarcoptes mites and the typical signs of sarcoptic mange are large areas of dry scabs and flakey skin which are irritating to the goat. As with all serious mite infestations the goat can lose condition and then may be susceptible to a secondary bacterial infection which may have more serious consequences than the original mange.

Ticks are not normally a problem in goats but it must be remembered that the sheep tick, which can carry the virus for Louping Ill, can also infest goats.[4] The virus, which can be fatal, can be transmitted to man through the milk and thus has serious implications if milking goats are kept in areas where sheep ticks are a problem.

There are a number of products that are effective for treating lice and mite infestation. Preparations containing bromocyclen ('Alugan', Hoechst Pharmaceuticals) are effective and also preparations that are simply poured onto the goat along its backline, such as 'Ovipore' produced by Rycovet Ltd, are convenient and effective, particularly against lice and ticks.

CLOSTRIDIAL DISEASE

The clostridium organisms are a group of bacteria which thrive in anaerobic (oxygen-free) conditions. They produce very powerful toxins and will be familiar more by the diseases some of them cause in man such as, tetanus, botulism and gangrene. Specific types exist in the gut of adult ruminants and normally do not cause problems. If, however, conditions in the gut become favourable they proliferate and the toxin they produce is usually fatal. The disease in adults is called enterotoxaemia and in kids pulpy kidney disease. Typical conditions when they become a problem is if the goats' feed is drastically changed resulting in a period of incomplete digestion before the rumen micro-organisms have had time to readjust.

There is an enormous range of vaccines available and all goats should be vaccinated against this disease. If in doubt about which to use veterinary advice should be sought. Normally goats are given an initial vaccination followed by another 4–6 weeks later and then a booster every 6 months. Kids from vaccinated mothers should receive their first vaccination at 10–12 weeks and those from unvaccinated mothers at 2–4 weeks of age. Pregnant goats should receive a booster 2–4 weeks before kidding as immunity will be passed on to the kids.

MASTITIS

Mastitis or infection of the udder should not be a major problem in a goat herd and if it is reasons should be sought. The time when the udder is most vulnerable to infection is immediately after milking and when young kids are still feeding from their mothers. The teat canal is relaxed and dilated after milking or suckling and infectious organisms may get into the udder at this time.

Because goats are such clean animals the udder does not get quite the same challenge from potentially infectious organisms as does the udder of a cow. Therefore, if a good standard of hygiene is practised at milking time mastitis should not be a problem. One of the most effective preventative measures is to dip the teats in a disinfectant immediately after milking. Proprietary teat dips are available, usually made up from an iodine based disinfectant and glycerine. A very cheap and effective dip is a 4 per cent hypochlorite solution which can be made up from a hypochlorite dairy disinfectant.

The first signs of mastitis are usually small clots in the milk. These may be seen if a few squirts of milk are drawn into a strip cup. This is a small vessel which is shaped to allow a black rubber disc to be fixed inside. When the milk is squirted onto this black disc any clots present are easily seen. These days small filters can be fitted in the milk lines of a machine milking system, as shown in Figure 7.3 and these also will show up any clots in the milk.

If the infection becomes more advanced the udder will become hot and inflamed. It may progress to a point where the udder, or parts of it, become cold. This is very serious as it means the udder tissue is ceasing to

Figure 7.3   Milking parlour with
mastitis detection filter arrowed.

function and could lead to the udder or at least half of it dying and
sloughing off. A generalised (systemic) infection may develop in which
case the goat may become very ill.

If mastitis is suspected a veterinarian should be consulted immediately.
A milk sample can be tested for infectious organisms and an appropriate
treatment plan put into action. Usually antibiotic creams are squeezed up
the teat canal into the udder and antibiotics may also be given by injection
to prevent the infection becoming systemic. It is most important that the
milk from goats that are receiving antibiotics should be discarded as
antibiotics can seriously affect the chances of making products such as
cheese and yoghurt.

If mastitis is a persistent problem all aspects of the goats' management
should be studied, particularly around milking, to see if a cause can be
found. Mastitis is more common in hand-milked goats than in well-
designed and well-run milking parlours. If new goats are being bought in
these may be the source of infection. Faulty milking machines could also
encourage predisposition to mastitis.

Johne's Disease

This disease, also kown as paratuberculosis, is also found in sheep and cattle but evidence suggests it may be species specific. In some experiments calves did not develop the disease after being injected with the organism that had been isolated from goats.[5] It is more likely that cattle could transmit the disease to goats. The bacteria damages the lining of the intestine resulting in a gradual wasting. This can take a long time to develop and during this time infected goats may pass on the bacteria without showing any signs of the disease.

Mercifully it is not common but those with infected herds will confirm what a problem it can be, with a few goats dying every year and with a drop in overall productivity. The tests for the disease are not very conclusive. However, if a goat looks a likely carrier from a skin and from a blood test and there is a history of the disease in the herd, it would be wise to cull that goat. A faecal test for the bacteria will confirm if the bacteria is present although it may take up to 6 months to get a result.

The bacteria can survive on the ground for a year so once the disease is present in a herd it is difficult to eradicate. Some success has been achieved in vaccinating against the disease and it is possible that, if kids from an infected herd are vaccinated and a rigorous testing and culling programme is practised, the disease could eventually be eliminated.

Tuberculosis (TB)

Although goats can become infected with bovine TB, naturally-occurring cases are virtually unknown and it is possible that they are not as susceptible as cattle. When the disease has occurred in goats it has nearly always been traced back to infected cattle.

Contagious Abortion (Brucellosis)

Although goats have been infected with this disease in experimental conditions no naturally-occurring cases have been recorded in Britain. The other brucella disease caused by *Brucella melitensis* is endemic in goats in Mediterranean countries and can cause the condition known as Malta Fever in man. Fortunately this disease does not occur in goats in the UK.

LISTERIOSIS

This disease is caused by the soil-borne bacteria *Listeria monocytogenes*. It has become more important in goats with the increase in commercial farming, particularly where goats are part of a larger farm enterprise, because the most likely source of the disease is silage. If silage is low-cut and, therefore, has soil in it and if the seal on the silage clamp or in the silage bales is not good the risk will be greater. The organism will not survive if the pH (acidity) of the silage is below 5.0, i.e. in conditions of high acidity.

Although occurring in three forms the commonest symptom is as a result of damage to the nervous system which causes the goat to circle round as though disorientated and it may stand with its head in a corner. Its temperature will be raised to about 40°C (104°F). These symptoms may last for 2 or 3 weeks or the disease may progress quickly with the goat dying within 2 or 3 days. It may also cause abortion or metritis in pregnant goats.

The organism can be shed in any body fluids from goats suffering from the disease and, therefore, these must be isolated. Some success has been recorded in treating the condition with the chlortetracycline antibiotics and during an outbreak of the disease all contact animals with a raised temperature should be isolated and treated.

This is a zoonotic disease − in other words it can be transmitted to humans − and is yet another example of the strong case for pasteurisation of all goat milk intended for sale for human consumption.

CAPRINE ARTHRITIS AND ENCEPHALITIS (CAE)

CAE is caused by a virus that is virtually identical to the *maedi visna* virus in sheep. It causes a slow loss of condition in goats even though there may be no loss of appetite. Eventually the leg joints become affected with considerable swelling and varying degrees of paralysis. In some countries the disease is known as 'big-knee'.

A survey demonstrated that of almost 3,000 goats tested throughout the UK 4.3 per cent were reactors (seropositive) to the test for this disease. This compares with an incidence of over 80 per cent reported in some countries. The goats that were seropositive represented 10.3 per cent of the 331 herds in the survey.[6] Because the incidence of the disease is low many goat-owners are keen to keep it that way if not eradicate it altogether. The Angora Society took the lead in this respect starting with the advantage of an imported breed where all goats had to be tested free of this disease before being allowed into Britain.

The Ministry of Agriculture now operate a Sheep and Goat Health Scheme under which they will test goats for a number of disease conditions, if the owner so wishes, including CAE. For a herd to be accredited as CAE free all goats must have had 2 clear tests 6 months apart and have not come into contact with non-accredited goats. Some people have their goats privately tested which is useful for the owner's peace of mind but is not recognised by the MAFF because only their scheme relies on documented proof of the restriction of movement of the goats.

METABOLIC DISEASES

Although pregnancy toxaemia, as sometimes seen in housed sheep, is not a particular problem in goats a condition related to it can be. If goats are too well fed they lay fat down in the abdomen. In some goats this can reach amazing proportions with a mass of fat weighing many kilograms. In Chapter 4 it was described how the goat's alimentary tract was modified for fibre digestion and how one of the main features was the very large rumen. The rumen and the other digestive organs take up a lot of room in the abdomen. In a pregnant goat the abdomen also has to accommodate, typically, anything from one to four kids. It will be fairly obvious if the abdomen is also full of large masses of fat that something must suffer.

When this situation arises the goat is not able to digest its food correctly and if short of energy will start to break down its own fat reserves. When this happens chemicals called ketone bodies are formed which poison the system and give rise to the condition called ketosis. One ketone is acetone and another name for the disease is acetonaemia. The acetone produced makes the goat's breath smell of pear drops, a diagnostic feature.

This is most commonly seen in overfat goats within the first week after kidding. The goat will appear listless and it will have lost its appetite (anorexia). Simple diagnostic test sticks are available which are tipped in a chemical which changes colour when it comes into contact with ketones. These can be used to test for ketones in milk, blood or urine.

Ketosis is a difficult condition to treat but an injection of corticosteroids and possibly a multi-vitamin injection may have a dramatic effect on about 50 per cent of cases. This is a condition where prevention is very much better than cure. Attention should be given to the goat's diet. During late pregnancy a concentrate providing adequate energy should be offered in 3–4 feeds a day.

There are, of course, many other disease conditions that can affect goats but these are covered in the veterinary literature much more thoroughly than is appropriate in this book. As suggested at the beginning of this chapter, goat farmers should work in close liaison with their vet, who hopefully will be a member of the Goat Veterinary Society, and should work out a health scheme together.

ZOONOTIC DISEASES

These are diseases that can be transmitted from animals to man. There are a number that may be found in farm animals, including goats, and some are common enough to justify special precautions. The important ones that all goat farmers should be aware of are:

1. Enzootic abortion – this can be a major problem in sheep in some areas.
2. Toxoplasma – carried by cats, also causes abortion in goats. Both of these diseases can cause abortion in women and, therefore, pregnant women or those who may be pregnant should never handle goats during the kidding period.
3. Ringworm – causes unpleasant sores on the skin.
4. Orf – causes unpleasant blister-like sores around the mouths of goats and sheep.
5. Anthrax.
6. Listeriosis – soil-borne bacteria, found particularly in silage of high pH (low acidity). May cause meningitis and abortion.
7. Louping ill – see details earlier in this chapter.
8. Foot-and-mouth disease.
9. Pox virus.
10. Brucellosis – contagious abortion in cattle, appears not to affect goats, causes undulant fever in man.
11. Leptospirosis – jaundice, carried by rats.

NOTIFIABLE DISEASES

There are some diseases that can affect goats where there is a requirement by law to inform the authorities of an outbreak or suspected outbreak. The authority can be the local MAFF Veterinary Department or the police. The notifiable diseases that can affect goats are:

1. Anthrax.
2. Brucellosis (*Brucella melitensis*).
3. Foot-and-mouth disease.
4. Tuberculosis.

MOVEMENT OF ANIMALS RECORDS

The Movement of Animals (Records) Order 1960 with amending Order 1961 requires all those keeping farm livestock to keep a written record of all movement of animals on or off their premises. These records allow the authorities to trace the likely route of a disease and animals that may be involved in the event of an outbreak of something serious like, for example, foot-and-mouth disease. Although any permanent record will do, books specially for the purpose are obtainable from HMSO. These books will be inspected several times a year.

HUMANE SLAUGHTER

Anyone involved with animals will, sooner or later, be faced with the problem of slaughtering. Usually this is because there is a sick animal that needs to be killed to alleviate suffering. For most situations it is preferable to call in the vet to do this, who may well be involved with the case anyway. However, there may be times when there is no alternative but for the farmer to carry out this task.

Adult goats should, preferably, be killed with a captive-bolt humane killer. The point of aim is at the back of the head just below a line between the base of the ears and towards the mouth. In an emergency a rifle or shotgun may be used but great care must be taken not to endanger the lives of people or other animals. If a farmer has no particular anti-foxhunting feelings the local hunt will often come and slaughter and take away animals as long as they are fit for the hounds to eat.

A veterinarian will kill a goat with an overdose of barbiturates injected into the jugular vein. In terms of minimising stress this is the preferred method but may not be possible in all situations, particularly in emergencies.

Kids can also be killed using a humane killer although this may seem rather drastic for very young animals. If large numbers of kids are to be killed, as may be the case with surplus males if no meat outlet is available, it may be worth while constructing a killing box. This is simply a reasonable air tight box large enough for a kid, which is attached to a pressurised gas cylinder containing carbon dioxide via a reducing valve. A small air bleed is fixed to the top of the box so that as the CO_2 is slowly fed into the box it will push air out through the bleed tap. If CO_2 is released into the box slowly the kid will drift quietly into unconsciousness and then the flow can be increased to fully charge the chamber. The gas is then turned off and the kid is removed some minutes later when there is no doubt that it is dead.

In the past many barbaric methods of destroying unwanted animals,

including kids, have been used such as drowning. This is quite unaccept-
able and it is important that farmers resolve the problem of humane
slaughter before the need arises.

REFERENCES

1. Farizy, P. (1970), 'Interent d'un traitment anthelmintique au thiabendazole
   chez la chevre en lactation', *Recl. Med. Vet. Alfort*, vol. 148.
2. Lloyd, Sheelagh (1982), 'Control of parasites in goats', *Goat Veterinary Society
   Journal*, vol. 2, no. 1.
3. Norton, C. C. (1986). 'Coccidia of the domestic goat, Capra Hircus, with
   notes on Eimeria ovinoidalis and E. Barkuensis (syn E. ovina) from the sheep!
   Ovis aries', *Parasitology*, vol. 92.
4. Reid, H. W. (1986), 'Ticks and tick-borne diseases', *Goat Veterinary Society
   Journal*, vol. 7, no. 2.
5. Collins, P., Davies D. C. and Matthews, P. R. J. (late) (1984), 'Mycobacterial
   infection in goats: diagnosis and pathogenicity of the organism', *British
   Veterinary Journal*, vol. 140.
6. Dawson, M. and Wildsmith, J. W. (1985), 'Sereological survey of lenti virus
   (maedi-visna/caprine) arthritis-encephalitis infection in British goat-herds',
   Veterinary Record, vol. 117, no. 86.

Chapter 8

Milking

The principal parts of a goat's udder, which are important to the milking process, are the teat canal through which the milk leaves the udder, the teat cistern into which the milk flows before it is removed from the udder and the glandular cistern into which the milk flows on leaving the glandular tissue where it is produced.

In cows the ejection of milk from the udder is an involuntary reflex action in which both the nervous and hormonal (endocrine) systems are involved. In the goat and sheep this reflex appears to be less critical and, therefore, the same degree of conditioned stimuli to achieve maximum let down and thus maximum milk yield does not seem to be so necessary. It possibly means that the milking routine can be a little more flexible although it is often reported that a change in the person milking can drastically reduce milk yields. So routines cannot be ignored even with goats.

Another difference between goats and cows is the fact that the cistern storage capacity is proportionally greater, which has implications for the timing of milking. In the cow, if the milking time is too uneven, the back pressure created by the build-up of milk may have an inhibitory effect on milk production and, therefore, yield. With a greater cistern capacity the milking interval is less critical in goats and experiments have shown that this would have to be more than 16 hours before a significant effect on yield would be recorded. This means goats can be milked at more civilised times than cows if the farmer so wishes!

Seemingly contrary to the evidence about the milking interval is information from experimental work that shows that if goats are milked 3 times a day the yield will be significantly increased.[1] This is most certainly the effect of the additional stimulation of the udder rather than anything to do with reducing udder pressure. It is well known that maiden goats will come into milk spontaneously and will produce quite impressive yields if they are suitably stimulated by milking.

The arguments for and against 3 times a day milking have to be examined carefully if this is being considered. It is pointless if the extra milking costs more in labour, effort and mechanical requirements, than the value of the extra milk that is produced.

Hand-Milking

Experimental studies of the natural sucking mechanism of goat kids, using radiopaque liquids and cineradiographs, have shown that the action of hand-milking is remarkably similar to the natural sucking mechanism.[2] To extract milk from the udder a kid compresses the neck of the teat between the tongue and hard palate and displaces the contents of the teat cistern into its mouth by compressing the filled teat also between the tongue and palate. The jaw is then lowered to allow the cistern to refill and the cycle is repeated. Creating a vacuum by sucking, while helping, is not essential.

Hand-milking uses exactly the same technique. The base of the teat is compressed between the forefinger and the thumb thus trapping the milk in the cistern which is then dispelled by applying pressure with the remaining three fingers in sequence, thus forcing the milk out of the teat. Experienced hand-milkers take a pride in milking quickly and to such good effect that with a good rhythmic two-handed action, they can milk a high yielding goat in 2–3 minutes and will have created a frothy 'head' on the milk several inches thick.

Goats can be hand milked almost anywhere but, in the interests of good hygiene, should always be milked well away from any bedding or bedded areas. A concrete apron kept clean for the purpose is the minimum required.

It is unlikely that anyone intending to farm goats on any scale would contemplate hand-milking except perhaps as an interim arrangement until some kind of milking machine was installed. Ten goats is about the maximum anyone should contemplate hand-milking twice a day seven days a week, assuming they all have good yields.

If goats have to be hand-milked the milking arrangement should either be so simple that nothing much is lost when the milking machine is eventually in use or a hand-milking area should be designed where milking machines can be used at a later date, with a minimum of structural alterations. For the former situation it is just a question of setting aside a clean area where goats can be tethered and milked. This system can be refined a little by having the goats on a simple raised standing with something to hold the milking bucket in place. It is quite convenient to use a wooden standing platform into which a hole can be cut to hold the bucket.

Whilst not essential it is a useful time to give the goats their ration of concentrate feed which will also encourage them to stand still for milking. To be able to do this some form of bucket or feed hopper system will be required and these can be attached to a simple yoke system that will hold the goats in position while they are being milked.

MACHINE MILKING

The first milking machine that depended on a vacuum for drawing milk out of a cow's udder was granted a British Patent in 1860.[3] It has only been during the last 20 years, with the increase in commercial goat farms, that we have seen much development and use of milking machines for goats. Much of the development of goat-milking machines took place in France where goat-milk production has, for many years, made a significant contribution to agricultural production.

All milking machines in common use today work on the same principle. They apply a pulsating vacuum to the goats' teats via a flexible liner fixed into a rigid case or shell. The liner is attached to rubber pipes which take the milk away to some kind of collecting vessel. The vacuum comes to that collecting vessel from a vacuum pump via an interceptor vessel which collects any liquid that may have got into the vacuum line. Also in the system will be a pulsator. This is a valve that alternately lets air and vacuum into the space between the liner and the shell. This causes the liner to open and close thus alternately squeezing and relaxing the teat as the vacuum draws milk out of the udder.

There are three main layouts or systems used and these are shown in Figure 8.1. The first layout using a bucket for collecting the milk is the simplest and cheapest – particularly when all of the equipment, including the vacuum pump, is mounted on a trolley as a self-contained unit that can be wheeled to where the goats are standing (Figure 8.1a). The layout in the drawing uses the same equipment but in this case it is partly fixed and partly mobile.

The second system shown is a direct-to-line system where the milk goes directly through the pipe line, to the bulk tank or other storage vessel, without going through any kind of recording system. The receiver jar in this drawing collects the milk from the goats. When the milk reaches a certain weight the jar will drop on its flexible mounting and switch on a milk pump which will pump the milk on down the line to the storage tank.

Finally, a similar system is shown but in this case there are recording jars which allow the milk produced by each goat to be measured before it is released from the jars and transferred along the pipes to the bulk tank. Recording systems using weight rather than volume and systems that measure volume with a flow meter are also used.

The Vacuum System

All machine-milking systems require a vacuum pump. This is usually driven by an electric motor but internal-combustion-engine-driven pumps are available if there is no electricity. The purpose of the vacuum pump is

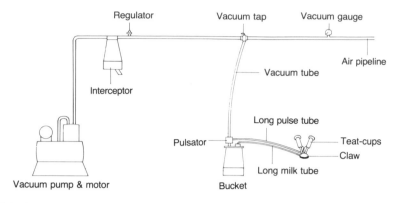

(a) A bucket milking system.

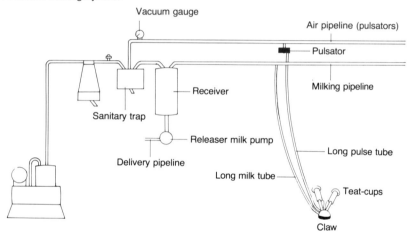

(b) A direct-to-line milking system.

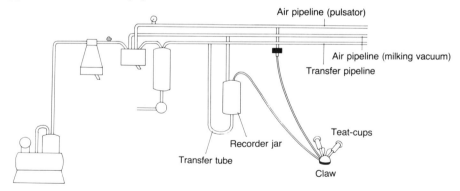

(c) A recorder jar parlour milking system.

Figure 8.1   Different milking parlour systems.

to remove air from the milking machine system thus creating and maintaining the vacuum necessary for milking. These pumps need to be kept lubricated and cool and, therefore, must be sited where regular maintenance is possible and where the air flow around the pump will not be restricted. Another important factor with respect to the siting of the pump is the exhaust. This can be moderately noisy even though it is normally run through a silencer similar to those used for motor car exhausts. It is preferable to point the exhaust away from walls or other surfaces that may reflect the noise.

The size of the pump will be dictated by the size of the milking system. The longer the pipes and the greater the number of milking units there are, the greater will be the requirement for vacuum. If the pump is not big enough teat cups will constantly be falling off the goats during milking.

Between the vacuum pump and the goats will be an interceptor vessel. The purpose of this is to intercept any liquid such as water and cleaning fluid and occasionally milk and to prevent it being drawn into the vacuum pump. Pipeline and recorder jar systems will also have a sanitary trap which separates the air system from those parts in contact with the milk, as shown in Figure 8.1b and c.

It is most important that the vacuum level in the system remains constant even though the demand will fluctuate as the milking units are put on and taken off the goats. To achieve this a regulator valve is fitted which opens and closes according to the level of vacuum in the pipes. A vacuum gauge is fitted so that the vacuum level can be checked at any time.

The Pulsation System

Pulsation is the opening and closing of the liners of the teat-cups which occurs when a valve mechanism or pulsator alternately connects the space between the liner and the rigid shell to the vacuum system and to the atmosphere. There are two types of pulsators in common use. The relay type is controlled from a central control unit and the self-contained type has a built in control mechanism. The latter is the favourite for the self-

Table 8.1 Mechanical settings for milking machines

	Cow	Goat	Sheep
Vacuum level (kPa)	44	37	44
Pulsation rate (ppm)	60	70–90	90–120
Pulsation ratio	50:50	50:50	50:50

contained bucket type milking units. Most modern central pulsator controls can be adjusted for a wide range of settings and can therefore cope with the different demands of cow, goat or sheep milking (Table 8.1).

Pulsationless Milking

From time to time the interest in pulsationless milking systems is revived and in recent years small units have been offered, particularly for the small-scale goat producer. The evidence for and against this type of machine is hard to find but there is a fairly strong body of opinion that suggests long-term use of continuous vacuum will result in teat damage. If a continuous vacuum is applied to a teat the blood flow through the skin capillaries is restricted until the vacuum is released and it is thought that this will lead to lesions on the teat.

The Milking Cluster

The cluster comprises the two teat-cups (each with a shell, liner, short milk and short pulse tube), a claw piece and the long milk and pulse tubes (Figure 8.2).

The teat-cup shells can be made from any rigid material but in practice are usually made from stainless steel, chromium plated steel or a tough plastic such as polycarbonate.

There is a bewildering range of teat-cup liners available for cows, with a variety of shapes and made from rubber of differing specifications. There

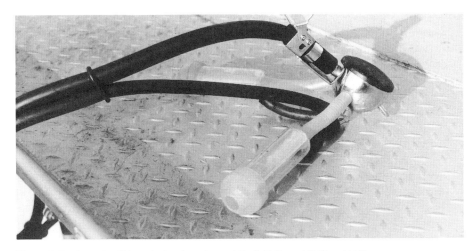

Figure 8.2   A goat milking cluster (courtesy of R. J. Fullwood & Bland Ltd).

is not such a wide range for goats although each manufacturer seems to produce its own particular design. Natural rubber absorbs fat and deteriorates fairly quickly and, therefore, synthetic materials or mixtures of synthetic and natural rubber are now used. Some of these, such as silicon rubber, are translucent and if used in conjunction with a transparent polycarbonate shell allow one to see the whole process of teat compression and milk expulsion. This may not be of great significance but it certainly adds to the interest of anyone, including the person milking, who is watching the process.

There is quite a lot of variation in the shape of the teats of most commercial dairy goats and as long as the teat-cup liner is big enough this should not be a problem. However, some units have been installed using the much shorter sheep teat-cups and it is likely, with these, that many of the goats' teats will be sucked against the base of the liner thus preventing the removal of the milk. The function of the claw piece is to receive the milk from the teat-cups via the short milk tubes and in its simplest form is just a sealed chamber connected to which are the short milk tubes delivering the milk and the long tubes removing it. The claw also has a small air inlet hole to facilitate the removal of milk under vacuum. Also attached to the claw piece will be the short and long vacuum tubes controlling the pulsation of the liner.

In recent years there has been a lot of research looking at ways of reducing the risk of mastitis and improving the hygienic quality of the milk by improving the efficiency of milk removal from the claw. In the latter respect some manufacturers have produced claws that have a shut-off valve that closes when the cluster falls off and the claw hits the ground. This will prevent contamination being sucked in and also will reduce the loss of vacuum.

There are also designs where ball valves are fitted to the inside of the claw piece which prevent blow back of milk up onto the teat and up the teat canal.[4] This is a problem when cups are removed or if they fall off, as the in-rush of air forces the milk back up the teat-cup. This problem is thought to be a cause of mastitis in cattle and it must be assumed that similar problems could occur with goats.

Milk Pumps

Pipeline and recorder systems will need some method of removing the milk from the milking machines to the storage vessel or bulk tank. With bucket units the bucket itself will be carried to the milk room or it will be emptied into a larger container such as a milk churn. In most pipeline systems the milk is collected in a receiver jar which may be fitted with a level switch or a weight switch that will actuate a milk pump. The weight switch operates when the milk, in the flexibly mounted receiver jar, weighs enough to move the jar down. As it does so it actuates the switch to the pump which pumps the milk to the milk room.

Milk pumps are either of centrifugal or diaphragm type. The former pumps the milk by means of a rotating blade and the latter by means of a rubber diaphragm moving in and out. The diaphragm pump will agitate the milk much less which may be important in some units (see the section on milk handling).

MILKING PARLOUR DESIGN

The time taken to milk a herd of goats can be easily doubled if the layout of the milking system is poorly designed. Anyone contemplating installation of a milking parlour should seek advice and should try to visit several goat units to see some different approaches to parlour design.

The first point to consider is the maximum number of goats ever likely to be milked. It is far better to design a parlour layout, however small, that is extendable if the need arises. This will be much less expensive than starting again with a new larger design.

Figure 8.3   A bucket milking unit (courtesy of R. J. Fullwood & Bland Ltd).

A Small-scale Parlour

The simplest approach is a small standing, with perhaps 3–6 goats standing at one time and milked via a bucket unit (Figure 8.3). This can be in
conjunction with a fixed vacuum pump and with a pipeline that allows the
bucket unit to be connected to the vacuum system at a number of convenient points. With this and most other layouts some form of feeding
yoke would be installed that would hold the goats in position during
milking and to which buckets or a feed trough could be attached. Milking
time provides a good opportunity to feed the goats their concentrate
ration. The complete goat standings can be mounted on a platform to
bring them up to a comfortable working height for the operator. The goats
would reach this platform by means of ramps or steps. Whatever is used it
is most important that the goats cannot slip. If they do so just once they
will be reluctant to use the steps or ramps again. The critical dimensions
for such a layout are shown in Figure 8.4.

However modest the design of a milking system or parlour it is very
important to install it in a building in such a way that the goats move in
and out quickly so that time will not be wasted at this point. The most
sophisticated and expensive installations can be wasted if the layout is
such that too much time is spent getting the goats in and out of the
parlour.

Figure 8.4   Dimension of steps for goats climbing on and off a milking platform.

If at all possible it is preferable to have the goats coming in at one end
and going out at the other. This way a nice flow can be achieved and, as
long as there are no 'bottlenecks' or hold-ups outside the parlour, the
movement of the goats should be accomplished at break-neck speed! It
should not be too much of a problem to modify a building to allow this
type of layout. Goats do not need large openings. A vertically sliding trap
door operated, from the milking pit, by a cord over a pulley is relatively
easy to construct, requiring little more than the removal of a few blocks or
bricks and a wooden frame put round the opening to take the sliding door.

Goats, being inquisitive animals, are always ready to rush off to some-where different. They also seem to enjoy being milked and therefore movement in and out of the parlour, through sliding trap doors, should not be a problem at all.

If there are more goats than will fit into the confines of the parlour it will be necessary to have a collecting pen outside the entrance and exit. It will then be possible to, virtually, run one group into the parlour as the previous group is going out. This will all help to speed up the milking operation. It is possible on a long-established farm that the goats will be so used to the milking routine that a collecting yard will be unnecessary. All that would be required would be some division, e.g. swing gates, that would prevent those to be milked from becoming mixed with those already milked. If this can be arranged it is then simply a matter of running the goats into the first half of the divided area and when the first batch have been milked these can be allowed to find their own way back to their pens. It helps, in such cases, if there is someone who can put forage in racks or feed into troughs, depending on the system, as this will give the goats the incentive to run back from the parlour without lingering and getting into mischief.

Larger Parlours

Although the principles already discussed would apply to any milking parlour, a small parlour with only 6 positions would not really be able to cope with more than about 50 milking goats. If a large commercial herd is to be milked a much larger parlour would be appropriate and as the size increases it becomes more relevant to consider the various layouts possible.

There are three very different layouts for goat-milking parlours and all have their advantages and disadvantages. The first system shown in Figure 8.5 is the side-by-side or abreast-type which, when the goats stand at an angle to the pit, becomes a herring-bone. Figure 8.6 shows a tunnel layout where the goats stand one behind the other. The illustration in Figure 8.7 shows a large rotary parlour in use in France.

With the abreast or herring-bone parlour there is also a choice of position of the goats. They can be facing towards or away from the operator. The herring-bone layout allows the operator to put on the teat-cups from the side and there will be little pulling of the teats in doing so. With the conventional abreast layout there is the choice of putting on the cluster from behind, through the back legs, or if the goat is facing the operator he/she will have to reach back to put the cluster on from the front.

Some people claim that if the goats are facing the operator they are less nervous and settle to the parlour routine more easily. Also, it is claimed

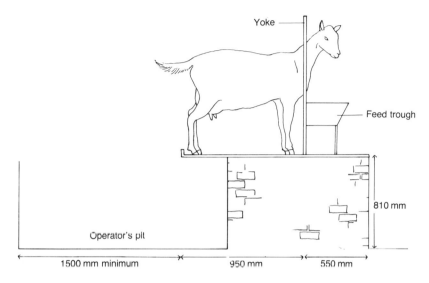

Figure 8.5 Dimensions of an abreast-type milking parlour.

that there is less risk of udder damage when the teat-cups are put on from the front. Probably the biggest advantage of the goats facing the operator is that feeding and locking and unlocking the yokes can be done from the pit.

The disadvantages of forward facing goats are that it is more difficult to inspect udders for problems and there is a tendency for the goats to drop food onto the operator and the milking equipment in the pit.

Modern developments have removed some of the problems. With various in-parlour systems and feeding yokes most of the operations involved in getting the goats into the parlour, feeding them and letting them out can be done by the person in the pit.

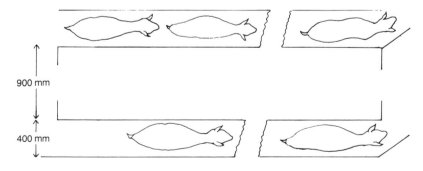

Figure 8.6 Layout and dimensions of a tunnel milking layout.

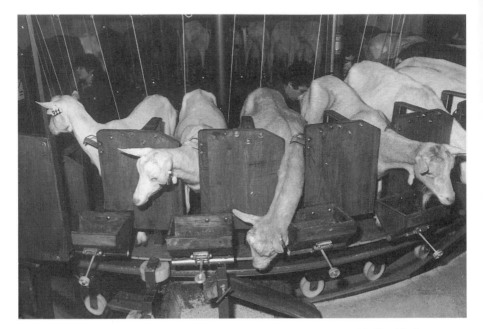

Figure 8.7   A rotary parlour in use in France (courtesy of R. J. Fullwood & Bland Ltd).

The majority of parlours installed in Britain are abreast types with the goats being milked through the back legs. No particular problems have arisen except where there have been faults in the installation or where the goats are particularly badly behaved. Figure 8.5 shows an abreast layout with the various critical dimensions.

The Tunnel Layout

This has the advantage of taking up less room than a side-by-side parlour and the throughput of goats will be faster. This is because the goats are not fed with this system. This may be seen as a disadvantage if individual feeding of concentrates is required. Some problems have been experienced with goats biting one another when in a tunnel parlour.

Rotary Parlours

Rotary parlours for cows seem to go in and out of fashion. One of the problems with cow rotaries is that they are fairly complex and there is a lot that can go wrong. There are two approaches to rotary parlours for goats. One is virtually a scaled-down version of the type used for cows and the other is a small hand-turned affair holding, perhaps, about 6 goats. The big advantage of rotaries is the fact that the throughput of goats will be much faster with one operator milking many more goats than with other systems. Some farmers will consider that this more than compensates for the possible increase in running costs due to the complexity of the system.

Moving Abreast

This is a French design that is manufactured and marketed in the UK by Fullwood & Bland of Ellesmere, Shropshire. With this type of parlour (Figure 8.8) the goats stand on a moving platform like a conveyer belt. As the 'belt' moves round, a feed yoke opens at the end where the goats are waiting to be milked. A goat puts its head into that open yoke and as the belt moves round, that yoke will shut and the one next to it will open, allowing another goat to step onto the platform and to put its head through the yoke to feed. After a few seconds the platform will be full, with all the goats fixed into the feeding yokes, and at which point it will be stopped. The goats are then milked in the conventional manner. When all of the goats have been milked the platform is switched on. As it moves the end yoke will open and a goat will be released. As this happens a yoke at the other end will open to let the first goat of the next group step on. By the time all of the first group have stepped off, a new group will be in place to be milked. As the platform moves round, feed hoppers at one end will

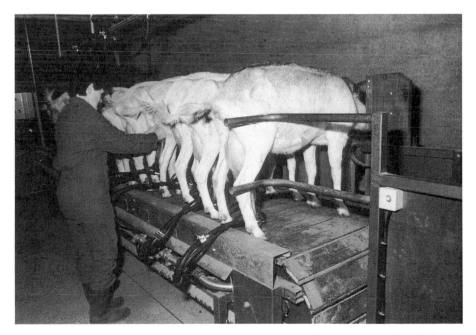

Figure 8.8   A 'Lactofeed' moving abreast parlour (courtesy of R. J. Fullwood & Bland Ltd).

tip up and waste food will be tipped out and when the empty hopper reaches the other end a feed dispenser will drop a predetermined amount of concentrate into the feeders. As with the rotary parlour this is a fairly complicated system but it can speed up the milking operation considerably. The manufacturers claim a possible throughput of 150 goats per hour with one operator.

THE INSTALLATION OF MILKING PARLOURS

It is important that the siting of a new milking parlour is carefully planned. The most important consideration is the movement of the goats in and out and some aspects of this have already been discussed earlier in the chapter. If at all possible, awkward turns at the entrance or exit of the parlour should be avoided.

A decision has to be made about whether to have the goat standing raised above floor level or whether to have an operators' pit. As goats readily climb, a raised standing may seem the best and certainly the cheapest solution. The digging of a pit would add considerably to the installation costs and there is often a problem with drainage.

There are, however, some problems with raised platforms that must be considered. One obvious one would be the height of the building or the height of any fittings that would be in reach of the goats when on the milking platform. Goats in this position have been known to strip fibre-glass insulation from water pipes. The dimensions of milking platforms are shown in Figure 8.6.

Even though, at present, there are no regulations concerning the standards necessary for goat-milk production it is advisable to construct and finish goat milking parlours in a way that would be acceptable under the Milk and Dairies Regulations which set the standards for cow milk production. The principal requirements are that the parlour should be washable and that clean water supplies are available.

To fulfil these requirements the surface of the goat-milking platform should be either smooth-finished concrete or metal, but not wood. Many of the milking systems that are available are fitted with a metal checker-plate base which is ideal. If a pit is to be constructed it should be finished with a washable material before all the milking equipment is installed. An impervious paint such as chlorinated rubber or a tiled finish would be suitable.

The siting of taps, electrical sockets, pulsator control, etc. should be carefully planned as they will all need to be readily accessible and if not, much time could be wasted. Similarly the in-parlour feeding arrangements need special thought. The ultimate in sophistication is a computerised system as used for cows where each animal is recognised by a device called a transponder, which would be attached to a neck collar. This would 'tell' the computer the identification number of the goat and the computer would then trigger off a feed dispenser that would drop an appropriate amount of feed into the goat's feed bin, according to her milk yield. Whilst such a system gives superb control over the ability to ration feed according to yield it is very expensive. It is doubtful if the cost would ever be recovered from any possible increase in milk yield.

Many farmers, especially when converting old cow parlours, have adapted existing cow feeders. These are often manual or semi-automatic and can usually be set to dispense a predetermined amount. On some farms one cow dispenser has been adapted to feed two goats simultaneously.

If feed dispensers are not available or are not to be fitted it is not too much of a chore for the milker to jump up and scoop feed, from a small bin on wheels, into the feeders.

Most parlour designs incorporate some form of feeding yoke which holds a goat in the milking/feeding position. Some designs work automatically so that as the goats put their heads into the yoke to feed from the feed hoppers the yoke closes, thus holding the goats in position until the milker is ready to release them.

If all the yokes are in the open position when the goats are let onto the milking platform there is a tendency for the first goat in to put its head into the first yoke, thus barring the way for the rest. This problem has been overcome by devising systems where only the furthest yoke will be open and when the first goat walks along the platform and puts its head in to that open yoke it triggers off a catch which allows the next yoke to open. Thus the goats are locked in in sequence from the furthest end. This system is often called a cascading yoke. With a little ingenuity it is usually possible to devise a method of setting and releasing the yokes from the pit, thus saving the operator's time.

MILKING ROUTINE

Whichever parlour system is used the milking routine will be more or less the same apart from the different ways of getting the goats in and out of the parlour. The first stage would be to release the goats from their pen and allow them to run into the collecting area. It helps if the milking machine is switched on at this stage as the noise will act as a signal to the goats and will make them even keener to run into the parlour. Once in the parlour they will lock themselves in by way of the yoking system. In tunnel-type parlours the goats may not be locked in but will be restrained by the goat in front.

Figure 8.9   A double-sided milking parlour with low-level milking units and recorder jars.

The feeders may have been filled before the goats entered if a flat rate feeding system is used or they may be fed once the goats are in and have been identified if a rationing system is used. Once the goats are fed the milking commences. It would be at this point that udders would be washed if that was part of the routine. Whether it should be will depend on the general standard of hygiene and the cleanliness of the goats. If the goats are not coming from a dirty environment, but are coming from a good clean straw bed or from pasture, it may not be necessary to wash the udders before milking. If, however, they are obviously dirty or if milk testing has shown that there is a problem with high bacterial cell-counts in the milk, udder washing is advised.

For udder washing to be effective it must be done with a disinfectant solution. Washing with plain water will almost certainly create more of a problem than not washing at all. A number of manufacturers produce a range of udder washes and many also supply spray systems that dispense the correct concentration of washing solution by way of a spray nozzle. If udders are washed they must each be dried with a separate disposable paper wipe or towel. If not dried a drip of washing solution, probably full of bacteria, will be sucked into the teat cups when they are applied. If a cloth is used instead of disposable wipes any bacteria present will be transferred from one goat to the next.

The teat cups are applied with the machine running but with the vacuum shut off at the cluster by means of a pinch-valve on the rubber milk tube or by the operator bending the tube as the cups are moved into position. As the cups are put up to the teats the vacuum is released and the teats are sucked into the cups. Care should be taken that the teats are not bent or pulled into an unnatural position as this can eventually result in damage. Most goat clusters have the cups fixed at an angle which corresponds with the forward facing angle of most goat teats.

It is a convention if mastitis is thought to be a potential problem to squirt a drop of milk into a strip-cup before the milking cups are put in place. This cup has a black rubber disc that will show up clots in the milk which may be the result of mastitis. Small filters fixed in a transparent holder are available to fix into the long milk tube (see Figure 7.3). These will catch any clots as the milk passes through and will be visible to the operator. If mastitis is suspected and the milk is being collected in a bucket system or is going through a recorder jar it should be discarded or used for animal feed until the infection has cleared up.

It usually takes about 2 minutes to milk out an average goat and when there is no milk to be seen flowing through the pipes the cluster can be removed. To do this the vacuum should be shut off as described earlier. If not there will be a blow back as air enters the system and teats can be damaged as the cups are removed or infectious organisms may be forced up into the teats.

If one operator is working with more than eight milking points, over-milking is bound to occur. This happens when the cups are still on the goat but no more milk is flowing. Experimental studies on cattle and experience with goats suggest that this is unlikely to cause any major problems but it is a waste of machinery and energy if this is regularly happening.

Once the teat cups have been taken off it is important to dip the teats with a disinfectant teat-dip solution as a precaution against mastitis. Immediately after milking, when the teat canal is dilated, the udder is vulnerable to infection.

If a direct-to-line milking system is used the milk will be propelled to the bulk or storage tank. If a bucket system is used the bucket will be tipped into the bulk tank after each milking or when it is full depending on whether yields are being recorded. With recorder jar systems the milk yield will be read and recorded and then a valve is moved to allow the vacuum to draw the milk into the receiver jar from where it will be pumped to the bulk tank.

When all of a group of goats have been milked, and their teats have been dipped, the yokes will be released and the goats will go out into a collecting yard or back to their pens depending on the system in operation. As they move out of the parlour the next group will be let in. When all of the goats have been milked the parlour will be washed down and the milking machine system will be washed through.

Cleaning the Milking Machine

There are three options available for cleaning milking equipment. Hand-cleaning with a detergent-cum-disinfectant is the only practical way of cleaning hand-milking buckets and utensils and for bucket machine systems. Recirculating disinfectant can be used for pipeline and recorder jar systems as can hot water rinse systems, usually referred to as ABW (acidified boiling water) systems.

With all systems the aim is the same. Milk residue must be removed and all milk contact surfaces should be cleaned thoroughly to remove any contaminating bacteria. The principles of cleaning to remove bacteria are the same for almost all impervious surfaces. First of all it is necessary to remove organic matter, in this case milk, which would inactivate any disinfectant and protect any bacteria. After this an appropriate disinfectant can be applied to kill any remaining bacteria.

For effective hand cleaning various items of equipment and services will be required:

1. A supply of hot and cold water and a length of hosepipe to fit the cold tap.
2. A wash trough large enough to take any of the large items such as the milking machine buckets.
3. A washable work surface or draining board (not wooden).
4. A selection of plastic buckets, brushes and dairy chemicals.
5. Somewhere to store the dairy chemicals.
6. At least one thermometer reading to 100°C.
7. Rubber gloves, a rubber or plastic apron and protective goggles.

The routine for cleaning, after milking, those surfaces that have been in contact with the milk should be, a rinse with cold water, a wash with a tepid solution of detergent–disinfectant and a final rinse with clean water to which hypochlorite may be added.

Immediately after milking the outside of the clusters should be washed and clean water should be drawn up through the cups and into the bucket by using the vacuum system. The unit can then be washed with the disinfectant–detergent solution.

After the final rinse it is very important to hang up the equipment or to place it on a rack to drain and dry. If left wet, residual bacteria may multiply and contaminate the milk at the next milking. Trials have been carried out to show the importance of the combined disinfectant–detergent wash.[5] There may be a considerable difference between the bacteriological quality of milk from equipment washed in detergent alone and that from equipment washed in a combined detergent–disinfectant.[6]

With the fixed milking equipment of a pipeline or recorder jar parlour system the cleaning has to be carried out with the equipment in place. There are two methods for doing this. The first, known as circulation cleaning, involves a water pre-rinse, a recirculated hot detergent–disinfectant wash and a final cold water rinse. Although much emphasis is usually put on the chemical part of this method it is the high temperature of the wash that is probably doing most good.

The second method is the ABW system which relies on the disinfecting effect of water at a much higher temperature than used for circulation cleaning. For this system a water heater is required that will generate enough water at a temperature of at least $77^{\circ}C$ to run through the system, over all the milk contact surfaces, for at least two minutes and preferably four. For any cleaning solution or hot water to be circulated through the system, jetter cups are fitted which plug into the teat liners and through which the various washes will flow.

It is good practice to check the effectiveness of cleaning systems regularly. In the case of circulation cleaning this can be done by disconnecting sections of pipe and visually inspecting for milk scale and by checking the microbiological contamination, if any, in the pipe. ABW systems can be checked by using temperature sensitive tapes on the various surfaces and recording whether disinfection temperatures are being achieved.

The real test will be the microbiological quality of the milk. A high bacterial cell count is indicative of contamination from the milking machine and the efficiency of the cleaning system should be suspected and checked. It is far better to be overzealous with cleaning and never to have a problem than to lose sales due to a reduction in the quality of the milk. Lost sales on these grounds are extremely difficult to win back.

References

1. Henderson, M. J., Blanchford, D. R. and Peaker, M. (1985), 'The effects of long-term three daily milking on milk secretion in the goat: evidence of mammary growth', *Quarterly Journal of Experimental Physiology*, vol. 70.
2. Cowie, A. T. (1979), 'Anatomy and physiology of the udder', *Machine Milking*, eds Thiel, C. C., and Dodd, F. H., Technical Bulletin No. 1 (The National Institute for Research in Dairying, Reading).

3. Hall, H. S. (1979), 'History and development', *Machine Milking*, eds Thiel, C. C. and Dodds, F. H., Technical Bulletin No. 1 (National Institute for Research in Dairying, Reading).
4. Griffin, T. K., Grindal, R. J., Staker, R. T., Shearni, M. F. A., Bramley, A. J., Simpkin, D. C., Higgs, T. M. and Westgorth, D. R. (1981), *Development and Evaluation of Control Techniques for Bovine Mastitis*, National Institute for Research in Dairying Annual Report.
5. Cousins, Christina M. (1981), 'Milk hygiene and milk taints', *Goat Veterinary Society Journal*, vol. 2, no. 1.
6. Mowlem, A. and McKinnon, C. H. (1983), *The Production of High Quality Goat Milk for Retail Sales*, Proceedings of the 3rd International Symposium on the Machine Milking of Small Ruminants (Vallodolid, Spain).

Chapter 9

Dairy Work and Milk Products

There are many textbooks on the processing of milk and many of these will be useful to any would-be goat milk processor. It is not, therefore, appropriate in a book of this nature to go into great detail about the manufacture of products giving methods and recipes. In this chapter an idea of what is involved will be given with the intention of making the faint-hearted think twice or to encourage those who are keen to have a go.

We are undoubtedly in an age where there is great interest in speciality foods, including dairy produce, and there is a lot of scope for those capable of adding value to their milk by processing it into cheese or yoghourt. In addition there is still a good market for goat milk for liquid consumption.

Milk processing is a big undertaking for those new to goats and although this may be the eventual aim many farmers start off content in being able to find an outlet for bulk milk with the minimum of processing.

THE DAIRY

Construction

Whatever the system or the size of the farm it will be necessary to have a dairy or milk room for, at the very least, holding the fresh milk until it is sold or processed. Ideally this room or building should be adjacent to, or not too far from, the parlour. If the goats are being milked by a bucket unit the buckets will have to be carried to the storage tank or churns and, therefore, the distance should not be far. If the milk is piped to the dairy there will be the danger of mechanical lipolysis if the pipes are too long or tortuous.

In terms of construction and finish the dairy should have an impervious interior with all surfaces, including the floor and ceiling, capable of withstanding regular washing and a high humidity. Blocks painted with an impervious paint or covered in tiles give the best wall finish, with smooth-finished concrete or quarry tiles for the floor. The floor should drain effectively with no low spots that may collect puddles.

A generous window is desirable as the environment should be light as well as clean. However, strong sunlight is undesirable as the dairy should be kept as cool as possible. Some form of sunshade would overcome this problem. Window sills and any other ledges should slope down so that they can be hosed clean regularly.

There should be at least one hot and cold water supply and it is useful if the cold water tap can be fitted with a hosepipe. A steam source is extremely useful, particularly if processing milk, and it may, therefore, be worth investigating the installation of a small steam boiler. Steam is very useful as a heat source and for sterilising equipment.

Ample power points will be required and these and all other electrical fittings such as switches and lights should be waterproof. An electronic fly killer is useful although it alone may not be capable of entirely controlling the fly problem on many farms. A fly curtain made up of closely hanging plastic strips across the doorway of the dairy will do much to keep flies out and if the windows can be opened wire fly screens should be fitted across these.

Equipment

The amount of equipment required will depend on the degree of milk processing intended. At the very least some means of filtering the milk from the parlour will be required along with some kind of storage vessel for the milk. A means of cooling the milk and keeping it cool will also be required.

If the milk is collected in buckets it will be filtered by tipping it through a disc filter. This is rather like a stainless funnel in which a fine cotton filter is sandwiched between two perforated metal discs. The cotton filter is renewed at each milking. There have been occasions when people have tried to economise by using some other filter medium. At one time nappy liners were used. This should never be done as many of these products are impregnated with chemicals and also the material may break up and particles may get into the milk.

In a pipeline system a 'sock' filter will be used. This is a sausage-shaped cotton filter that fits over a perforated core inside a special tubular holder on the end of the milk pipeline from the parlour.

The problem of lipolysis or breakdown of the fat due to enzyme action and the need for efficient milk cooling has already been discussed. The simplest effective cooling method is the in-churn cooler. This device is connected to a cold water tap by way of a short hosepipe. The cold water runs through a rotating tubular paddle and then comes out with some force through small jets. This force turns the paddle, and the water is deflected to run down the outside of the churn. Thus the water on the outside of the churn and in the paddle cools the milk and the paddle ensures that the cooling effect reaches all milk in the churn. The obvious problem with such a system is that during warm weather the mains water may not be very cold.

If milk is stored in large quantities a bulk tank will be required. With some difficulty small bulk tanks of around 200 litre capacity can be obtained but most tanks from cow dairies will hold several thousand litres. A bulk tank consists of an insulated stainless steel tank which will have cooling elements from a built-in refrigeration unit around it. Attached to the lid will be an electrically-driven agitator paddle which will ensure the cool milk will be continually mixed. Most bulk tanks will cool the milk to and hold it at 5°C. Mobile bulk tanks are available and one of these may be suitable for a producer who delivers bulk milk into a processing plant.

Table 9.1 The effect of temperature on bacterial growth in milk stored for 24 hours

°C	Colony counts/ml
0	2,400
4	2,500
5	2,600
6	3,100
10	11,600
13	18,000
16	180,000
20	450,000
30	1,400,000,000
35	23,000,000,000

Apart from inhibiting enzyme lipolysis, cooling is most important to slow down the multiplication of bacteria in the milk. With an efficient cooling system, milk of good bacteriological quality can be stored for a number of days. Table 9.1 shows the effect of temperature on bacterial growth in raw farm milk.

To ease the cooling load of the refrigerated bulk tank or to increase the efficiency of cooling, a plate cooler may be used to cool the milk as it goes into the bulk tank. This is made up of a series of thin metal plates which, when sandwiched together, form thin hollows. These hollow sections are alternately connected to the milk and a cold water supply so that the milk flows between the sections through which the water is flowing. These plate coolers are very efficient because of the large area of the cooling surfaces. If necessary the cooling capacity can be increased simply by increasing the number of plates. Like the churn cooler, however, the efficiency of the plate cooler will be affected by the temperature of the mains water. If this is a problem a double-banked plate cooler can be used. With these one half is cooled by mains water and the other by water from a chiller unit. A double-banked plate cooler, of adequate size, will be able to bring milk at pasteurisation temperature of 72.5°C down to 5°C.

Milk of particularly good microbiological quality that has been efficiently cooled and stored below 5°C will keep for several days prior to processing. Regular bacteriological tests should be carried out, however, to check the quality of the milk because even the best systems can break down and contamination of milk can often occur from the most unexpected sources. It is much better if problems are discovered on the farm rather than by one's customers.

METHODS OF PROLONGING THE SHELF-LIFE OF GOAT MILK

Freezing

One useful way of prolonging the shelf-life of goat milk is to deep freeze it. If frozen correctly, unlike cow milk, it does not separate on thawing. Goat milk is often frozen for retail sales and for storage prior to processing into cheese, yoghourt or ice cream.

The most critical factor for deep freezing goat milk seems to be the actual storage temperature. It is vital that the temperature of frozen milk does not rise above −18°C. If it does separation of the fat is likely to occur. The sort of situation when this can be a problem is when liquid milk is put into a freezer where some milk is already stored. When the liquid milk goes in the temperature rises and as most domestic freezers run between −18°C and −20°C, this is likely to be well above −18°C. To overcome this a separate freezer should be used for freezing. Some producers use a blast freezer which will give very rapid freezing, but freezing time, as long as it is within 12 hours or so, does not seem to be nearly as critical as the storage temperature. Walk-in freezers with a very large capacity will probably cope with both jobs.

Problems may also occur when frozen milk is purchased and moved to another freezer for storage. If the time interval is too great the fluctuations in the temperature of the milk may result in the same separation problem.

It is important when freezing milk that it is not placed into the deep freeze in large volumes without a good air space around each pack or container. If frozen in plastic containers these should be stacked like children's building blocks with a gap between each. If the milk is frozen in polythene sachets or bags these should be put in single layers on trays or in wire baskets. Once completely frozen they can then be stacked more closely in the storage freezer.

Pasteurisation

Pasteurisation is the process of heating milk to destroy bacteria and thus reduce the risk of transmission of disease. There are two principal methods: one involves heating the milk to 63°C for 30 minutes and the other, called high-temperature short-time, uses a temperature of 72.5°C for 15 seconds.

There is much debate and discussion about the need for and the desirability of pasteurisation. Some people suggest that pasteurisation destroys the natural goodness and nutritional value of the milk. Certainly it will reduce the vitamin content of the milk and it will destroy some of the natural enzymes, the latter being an advantage as this will reduce the problems of taint discussed in Chapter 8. These relatively minor changes to the nutritional value of the milk have to be balanced against the risk of the disease-causing (pathogenic) micro-organisms carried by milk. Goats like many other animals can transmit some diseases to humans. Some that may be transmitted in milk are listed in Table 9.2. In addition to these pathogenic micro-organisms pasteurisation will also destroy milk souring bacteria and will thus improve its keeping qualities. It must be remembered that because goat milk is so often consumed by those who already have a health problem, the case for pasteurisation is strengthened.

Table 9.2   Some diseases that can be transmitted to humans from milk

Salmonellosis	Louping Ill
Tuberculosis	Toxoplasmosis
Brucellosis	Streptococcal infections
Listeriosis	Staphylococcal infections
Q Fever	Campylobacter infections

Milk can be pasteurised satisfactorily using any method of indirect heating. The simplest system, which could be used for small quantities in the home, would be to heat the milk in a small pan which would be placed inside another pan containing water. As the water heated up so would the milk and with the aid of a clock and a thermometer, it could be held at the correct temperature for the correct time.

Commercial-scale batch or holder pasteurisers work by the same principle. These will comprise an inner stainless steel tank surrounded by a steel jacket. In the space between the two will be the heating source which is usually either steam or water. If water, this will be heated by an immersion heater. A paddle is fixed to the lid and this will agitate the milk to ensure even heating throughout. Until recently small-batch pasteurisers were hard to obtain but ones of 110–220 litres are now available and would be suitable for some goat enterprises.

Figure 9.1   A 200-litre batch pasteuriser (courtesy of F.
Read & Son Ltd, Wilmslow).

There are two options with regard to the temperatures used. When pasteurisation was first used the main concern was to kill the tuberculosis bacteria and to do this it was necessary to heat the milk to 63°C and to hold it at this temperature for 30 minutes. As it happened most other micro-organisms are killed at this temperature. More recently a high temperature short time (HTST) method has been developed where the milk is heated to 72°C for 15 seconds. Either method will kill about 99 per cent of the bacteria in the milk.

Some of those who argue against pasteurisation voice concern about changes to the flavour of the milk. Extensive work with taste panels has not shown any problem with flavour in pasteurised cow milk and it is likely that the same would be concluded about goat milk, as long as it was of good quality to start with.

Figure 9.2 A continuous-flow high-temperature short-time pasteuriser (courtesy of F. Read & Son Ltd, Wilmslow).

An alternative type of pasteuriser is the continuous-flow type where milk flows in one end and comes out pasteurised, and in some cases recooled, at the other. These work on the higher-temperature short-time principle and are designed very much like the plate coolers described earlier. The milk flows through a system of plates which are heated by steam or hot water (Figure 9.2). As with the plate coolers this system allows rapid heat exchange because of the large surface areas involved. The efficiency of this type of pasteuriser is governed by the number of plates there are in proportion to the volume of milk flowing through.

Continuous flow pasteurisers are more expensive than equivalent batch pasteurisers but are considered to be more efficient. Certainly those that incorporate a plate cooler are very useful when used in conjunction with some form of continuous processing system such as bottling or cartoning. The smallest of this type of pasteuriser will have a throughput of about 50 gallons an hour.

Ultra High Temperature (UHT) Goat Milk

Cow milk has been successfully treated, by heating it to 140–150°C, to kill virtually all micro-organisms and, therefore, considerably increasing its keeping qualities to give a shelf-life of 6 months or more. This process, known as UHT, would seem to be very useful for goat milk, bearing in mind the widespread demand for medical patients who may have difficulty in finding a local supply.

So far problems have been experienced using this technique for goat milk, particularly with respect to possible seasonal changes in the milk causing precipitation or sedimentation after the heating process. Anyone wishing to develop such a product is advised to try out the technique on a number of small batches before risking large quantities or perhaps more importantly large amounts of capital for equipment.

Packaging Liquid Milk

The sale of milk for liquid consumption is attractive, particularly for enterprises that are early in their development, because the returns are good and no particular skills are required. However, packaging the milk from small enterprises for retail sales can be costly. There are four main choices of package that might be considered by any serious producer. These are polythene bags or sachets, cartons or rigid bottles made from plastic or glass.

Polythene sachets are relatively cheap, they can be obtained pre-printed with a producer's own label and there are machines that will fill and seal them automatically. They do, however, have disadvantages. Many outlets, particularly supermarkets, do not like this form of packaging as it is difficult to make a sachet look attractive on a shop shelf. They are used particularly for frozen milk but the disadvantage then is that if a hole is made in the sachet or bag this does not become apparent until the milk has thawed out when it will then leak out.

Cartons are an attractive way of packaging milk but they, and the equipment to fill and seal them, can be expensive. Cartons are available made from waxed cardboard or plastic. In the case of waxed card they are available either as a preformed carton with a closed solid base or as an open tube that has to be formed and the bottom sealed before any milk can be put in.

Preformed cartons, whether cardboard or plastic, are expensive but so is the machinery for forming and sealing the open type. Unfortunately most of the equipment developed for carton filling and sealing is designed for large-scale operations and the limited amount of equipment that is available for smaller enterprises is usually slow to operate and expensive.

Cartons can be filled and sealed by hand but this, of course, is extremely time consuming. Whatever method is used the filled cartons should be heat sealed. Plastic or metal clips are available for holding cartons closed, but as these do not form an air tight seal they are not suitable for milk that is expected to remain wholesome for a number of days.

Disposable plastic bottles, sealed with a foil top, are an attractive alternative to cartons. Small machines are available for filling and sealing these and for some, the manufacturers claim a throughput of 400 bottles an hour.

Dried Goat's Milk

Another possibility for extending the shelf-life of goat's milk is to dry it to produce a powder. As with all other processes it is important that the milk to be dried is of good microbiological quality. If it is dried, milk will keep for 6 months or so with the added advantage, when being sent around the country, of not containing water which adds to the weight and, therefore, cost of transport.

There are a number of processes used for drying milk, each requiring considerable expertise and specialist equipment. One of the oldest, and still used, methods is to run a film of milk over a heated drum which will cause the water to evaporate leaving just the solids. These are then scraped off and sieved to create a powder.

A more common method used these days is spray drying. This involves forcing pressurised, condensed milk through sprays or atomizers into a stream of hot pre-cleaned air. The dried milk is then removed from the air chamber, dried some more, cooled, pulverised and sieved.

Apart from being reconstituted to give liquid milk, powder such as this is useful for producers of yoghourt, etc. when there is a need to thicken the product. Before dried goat's milk became available this was always a problem especially as cow milk powder cannot be used for products intended for allergy sufferers.

GOAT'S MILK PRODUCTS

Cream

The proportion of small fat globules in goat's milk is smaller than in cow's milk. This results in the fat remaining evenly distributed throughout the milk for longer and consequently it takes longer for a cream layer to form. This also means that it is difficult to separate the cream. However, most good cream separators adjusted to a fine setting should give adequate results although it should be noted that the quantity and characteristics of the fat will change at different stages of the lactation period.

Butter

Very few people make goat's milk butter and certainly not in large quantities. As those who have tasted it extol its flavour in comparison with cow's milk butter it may be that a good market for it could be developed. The problem of separating goat's milk cream is the main reason for the small interest in butter manufacture.

Butter is made by separating the cream and then agitating it in a specially made, rotating butter churn. Small-scale hand operated churns are available but for large commercial scale enterprises they would be driven by powerful electric motors. Water is gradually added to the cream as the butter granules form. Once these granules have reached the correct size they are washed with more water and then they are gathered into one lump. After further washing this is worked with flat wooden pats into whatever shape is required. It is then left to cool for at least 12 hours.

If salted butter is required the salt is added to the final rinse before the granules are collected together. Goat's milk butter is white and sometimes colour is added if it is to sell alongside cow's milk butter.

Ice-Cream

Goat's milk ice-cream is particularly delicious and should be considered as another attractive way of adding value to the milk. Relatively small-scale ice-cream making equipment is available thanks to the demand from the many small family ice-cream businesses.

Yoghourt

The first yoghourt was made, many centuries ago, by peasants in Balkan countries as a means of increasing the life of the milk from their goats. Without understanding the chemistry involved they used bacteria to alter the characteristics of the milk to improve its keeping qualities. These bacteria break down the lactose or milk sugar to give lactic acid. The lactic acid acts as a preservative and prevents the deterioration or souring of the milk. The bacteria used these days for yoghourt production are *Lactobacillus bulgaricus* and *Streptobacillus thermophilus*.

The manufacture of a commercial yoghourt for eventual retail sale requires a clean environment, a reliable starter culture and milk of known good quality. To achieve a reliable product the milk would normally be pasteurised before processing.

Figure 9.3   A range of goat dairy products from Lower Basing Farm, East Sussex.

Yoghourt is produced by adding a starter culture to warm pasteurised milk and then leaving this to incubate at a temperature of between 37.5°C and 45°C. The starter culture is most important as the proportions of the bacteria must be correct or otherwise the flavour of the yoghourt will vary.

Incubation can be carried out in the yoghourt pots in a specially made incubator cabinet or it can be done in bulk even in the pasteuriser vessel, as long as the temperature can be regulated. If correctly incubated the yoghourt will set in 6–8 hours after which it can be transferred to a refrigerator.

Goat's milk yoghourts tend to be very liquid but it is now possible to thicken them with powdered goat's milk. Before this became available the preparation of pure goat's milk products for people intolerant to cow's milk was a problem, as dried cow's milk was the only available thickening agent.

Flavourings can be added and producers should note that the legal requirements for labelling such products are very specific and are ever changing. Expert advice from packaging companies or consultants should be sought before expensively labelled pots are purchased.

Cheese

The most organised goat's cheese producing country is France where the milk from about 1.2 million goats is used to manufacture about 40,000 tonnes of cheese each year. These cheeses are many and varied according to the method of manufacture and the region where they are made. Many people spending holidays in France will have come across cheeses such as Saint-Maure, Valencay, Crottin, and Chevrotin. It is perhaps encouraging for the British cheese producer, however, to learn that, for the last 2 years, top honours at the International Cheese Show at Nantwich have gone to British produced goat's cheese.

Cheeses can be grouped into five main types, fresh, soft, blue, hard and whey, according to the way they are made and their subsequent characteristics. Many of the French cheeses are made from unpasteurised milk which, owing to the range of micro-organisms present, adds to the variety of the characteristics of the cheese. French farms producing cheese from unpasteurised milk do, however, have to be certified as clean milk producers and that they are free of brucellosis, a problem in Mediterranean countries.

Cheese is made by causing the protein, or more specifically the casein, in the milk to coagulate. The casein occurs as small particles called mycelles and it is these that come together to form a more solid mass. This coagulation is achieved by the action of lactic acid and rennet. The degree of one type of coagulation over the other will dictate the type of cheese produced.

Lactic-acid coagulation is achieved by allowing the natural lactobacilli, or if the milk is pasteurised, an added starter culture, to break down the lactose to form lactic acid. Other bacteria in the milk and the temperature will influence the characteristics of the lactic curd. It will always be soft, delicate with a high moisture content.

Rennet, used for the manufacture of cheese, traditionally comes from the stomachs of calves although other types are now available. Only those produced by a reputable supplier should be used if disappointing results are to be avoided. Rennet causes the casein mycelles to floculate forming a net-like mass which catches most of the other solids in the milk.

Fresh cheese is produced after a lactic curdling time of up to 24 hours. The curds are drained for a short time and then put into a mould and the resulting cheese is usually eaten after only a few hours.

Soft cheese is produced in a similar way but it is drained for much longer to separate the whey. After draining it is left to ripen which is brought about by enzymes in the milk, rennet and also by the bacteria which develop in the curd. Ripening may take from 5 to 30 days and the characteristics of the final cheese will depend on the ripening time and the bacteria in the curds. Salt may also be added to the curds.

Blue cheese is prepared from curd produced by both lactic acid and rennet which, after being cut into blocks, is inoculated with the blue mould *Penicillum glaucum*. The curds are ripened in a controlled environ-

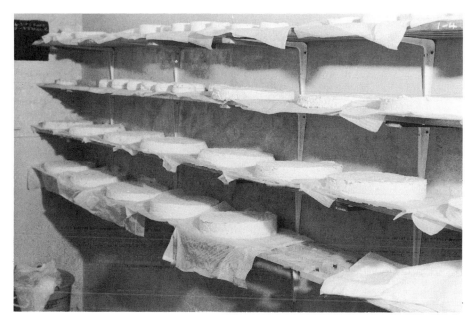

Figure 9.4 Cheeses in the ripening room.

ment where a temperature of 9–10°C and a relative humidity of 90–95 per cent can be maintained throughout the ripening period of up to 5 months.

Along with blue goat's cheese, hard cheese is in shortest supply because few people have the correct environment or the necessary skills to produce it. After draining, the curds are put into moulds and are pressed in special cheese presses to squeeze out most of the moisture. The cheeses remain in the press for up to a few days depending on the type. They are then turned out of the moulds, possibly salted, and wrapped in a cheesecloth. The maturing or ripening time will depend on the type of cheese. Most cheeses will be ripened for several months in a dark, well-insulated room, maintained at a temperature of 8–10°C and with a relative humidity of 80–90 per cent. An old farmhouse cellar may provide almost ideal conditions. It is important that the cheese room is well ventilated to remove heat from the ripening cheeses and to keep the temperature even throughout the room.

MARKETING GOAT'S MILK AND ITS PRODUCTS

The increase in the demand for speciality foods has created a good market for good goat's milk products. A few years ago it would have been hard to believe that some of our major supermarket chains would be stocking goat's milk, cheese and yoghourt.

The key words for successful marketing are the same for any product, including all of those from goats. These are quality, quantity and continuity. The image that the product conveys may also be important.

The quality of any food product must be first class. If it is not, people will not buy it a second time. Also goat producers should always bear in mind the fact that many of the consumers of their products will already have a health problem and it would be unfair to supply them with anything but top quality products.

It takes a long time to build up a reputation for supplying good products but it may take just one batch of products of inferior quality to destroy the market forever.

It can also be good for sales if the customer is able to see how the product is manufactured. This is almost mandatory if one is selling through a dairy buyer for a chain of stores. If conditions can always pass such an inspection the risk of poor hygiene affecting quality will also be considerably reduced.

It is important that once market outlets are found, production can keep up with the demand. Failure to deliver the quantities agreed may destroy a market almost as quickly as a reduction in quality. This can be a particular problem for goat producers whose milk production is geared to the natural season. The shortage of milk during the winter months has been the main reason for many goat producers finding difficulty in selling their milk. Ways of achieving a more even all-the-year round production are discussed in Chapter 5.

To achieve continuity means producing both quality and quantity consistently. A customer will expect products to be of the same high standard and with no variation within the type.

Creating the appropriate image to sell a product is much more difficult because different images influence different people in many ways. Goat producers are at an advantage in having an image already. Goats are associated with the country, with less intensive agriculture, with health, in fact all the things that most people perceive as being missing from modern agricultural production.

An image is created in a number of ways, by the farmer him/herself, by the appearance and location of the farm and by the appearance of the product. Until recently most people's image of a goat-keeper was that of elderly eccentrics who spent most of their time attending goat shows and scouring the hedgerows for browse or forage for their goats. The goat itself conjured up a smelly, independent, wild-spirited animal with a reputation for eating anything.

To some extent, thanks to the more serious press coverage given to goats and goat farming, things have improved. Many people are aware that goats can be farmed in large numbers, many will have tried goat products when on holiday abroad and many will know someone who keeps goats or will have at least seen some in a farm park or similar tourist attraction. This latter point now means many people will have discovered that, with the exception of the entire male, goats do not smell as much as most other farm animals.

Goat producers are now, by and large, seen as normal farmers. It is important for a goat producer to always look tidy and business-like and to always be business-like and take note of a customer's requirements. As goat producers work with little bought-in labour they tend to be out of their offices for long periods, so that a telephone answering machine is essential. It is annoying and frustrating not to get a reply when telephoning but being able to leave a recorded message is a compromise that most people will accept.

Packaging is very important in conveying an image and, therefore, in selling a product. All packaging must protect the product and should always look clean on the outside.

Choosing a colour and a label design is such an important selling point that it is probably worth seeking the advice of an expert. Certain colours are better for selling certain products. An obvious example is that bright harsh colours do not sell dairy products.

A picture of a goat or even a goat's head will make a packet more attractive, apart from immediately telling the customer that it is a goat product. Country scenes always help sell dairy products. It must always be remembered that the packet must look good from the minute it leaves the production line to the time the customer buys it. This means the packet must be able to withstand a fair bit of wear and tear. Damaged packets definitely inhibit sales.

White is a difficult colour because it soon looks dirty and dirty packets definitely also inhibit sales. Some colours such as black are never associated with milk products but it may be that the first person to try it will find that for some inexplicable reason it captures the imagination of the public. There have been many instances of products being sold by their packaging rather than anything to do with the product it contains.

In addition to attractive packaging the name is also most important in creating the right image and in giving the product an identity. A good example of this is the Milk Marketing Board's Lymeswold cheese. Like country scenes on packages, country names are useful aids to selling. Most goat farms will be in the country. Is there a local name that can be used? It can be regarded as a bonus if the name gives the product a regional identity. If there is not a suitable local name to use then, as with Lymeswold, make one up.

From all this it can be concluded that to be successful in marketing dairy goat products one must be business-like in one's approach, the quality of the product must be first class and it must be available all the year round in the quantities that have been agreed with the customer. In addition the product must be attractively wrapped or packaged and be given a name that helps create the correct image and which gives the product its own special identity. It may be that for this aspect, in particular, money spent on professional advice would be money very well spent.

REFERENCE

1. Davies, J. G. (1955), *A Dictionary of Dairying* (Leonard Hill, London).

Chapter 10

Fibre Production

Goat's hair has probably been used for making cloth for virtually as long as goats have been domesticated. In the Bible (Exodus 26) Moses is ordered to make curtains of goat's hair to cover the Tabernacle; it is likely this was mohair. Today goat's fibre is used for a variety of applications from the production of fine cloth and yarns for high fashion through to carpets and paint rollers.

GOAT'S FIBRE

Most goats have two types of hair growing from primary and secondary follicles. In most breeds the primary hairs make up the main coat and in some this is very long giving the goats a distinctly shaggy look. Generally speaking thick long coats are usually found on goats in colder environments. Some breeds of goat, again usually those in cold environments, produce soft, insulating, down-like secondary fibres. This down fibre is cashmere and is highly prized as a fashion fibre, resulting in goats being selectively bred for its production. In the textile trade the name cashmere is only used for goat underdown that is less than 19 thousands of a millimetre (microns) in diameter.

Angora goats have been selectively bred, for centuries, for their secondary hair growth and they now have a dense, long, fine coat of this secondary hair which is called mohair. Mohair ranges in diameter from 23 to 38 microns depending, mainly, on the age of the goat. The primary hairs in an Angora's fleece, left over from its double coated ancestors, are called kemp and are undesirable, as they will adversely affect the quality of the finished cloth.

A particular characteristic of mohair is its lustre. All hair has scales on the outer surface of each hair fibre. These scales on mohair lie almost flat giving good light reflection which accounts for the lustre or shine. At its best this lustre will make an Angora look almost silvery in a favourable light.

The increased interest in Angoras, particularly in countries that have to import them, has resulted in a lot of cross breeding towards grading-up to

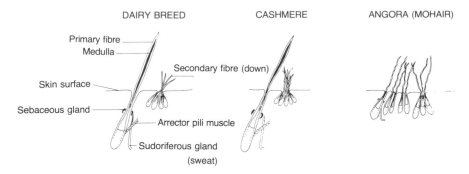

Figure 10.1 A comparison of goat fibre types.

pure bred status. This cross breeding has produced a third, and entirely new, fibre type which, in characteristics, falls between cashmere and mohair sharing some qualities with each. This new fibre has been called cashgora. It is under 22 microns in diameter and usually has a slight lustre, which with its fineness makes it very soft and silky to the touch. Manufacturers have welcomed this new fibre even though it is not yet being produced in commercial quantities.

HUSBANDRY OF FIBRE GOATS

In general the husbandry of fibre goats will be a mixture of sheep and goat farming practices. Angora and Angora cross-bred goats in many ways compare with lowland sheep and cashmere goats compare with hill sheep.

Housing

From the point of view of financial return or from the nature of the product it is not appropriate to keep fibre goats in the same intensive systems, with a high requirement for labour, as one would with dairy goats. Much better use would be made of natural grazing with the goats, depending on the weather, spending most of the time outside.

We are still learning about this aspect of goat farming and over the last few years have come to understand that even pure-bred Angora goats are, like most other breeds, adaptable and seem to thrive in environments very different to those in their country of origin. Also the grease content of fleeces varies a good deal and it is likely that greasier fleeces will be more waterproof. If the land remains dry Angora goats can remain outside for most of the year. Field shelters, that provide protection against incessant rain, would be required particularly during the colder periods. During the summer natural protection from trees or rocky overhangs would suffice.

Field shelters should be large enough to hold several goats and ideally should have either wide or several openings to stop bossy goats preventing others from sheltering. Ideally field shelters should be mounted on skids and light enough to be pulled to a fresh position when the ground around them becomes muddy. Slatted floors in shelters will help prevent fleeces becoming badly soiled when the goats lie down. Slats made from 5 × 2.5 cm softwood with gaps of about 1.5 cm seem to be the most suitable. The slats should be well supported from underneath because if they break injuries are likely. (See Figure 3.9 in Chapter 3).

Cashmere-type goats will be hardier than Angoras but this will reflect more on where they are kept rather than changes in husbandry. The type of cashmere goats we have in Britain, at the present time, have low yields of cashmere fibre and would be hard to justify in a lowland situation. However, in terms of output they compare favourably with hill sheep and it would be in a hill-type environment where they would most appropriately be farmed. They also would require shelter if natural features did not provide it.

With both Angora and cashmere goats the provision of shelter after the fibre has been harvested is important. Both may require a period of housing immediately after shearing, particularly if temperatures are low, if cold shock is to be avoided.

Feeding

The nutrition and feeding of fibre goats will be more or less the same as for dairy goats bearing in mind the greater use of natural grazing and differences in productivity between the breeds. Fibre is a proteinaceous material and, therefore, a fast rate of hair growth, as seen in Angoras, will require an adequate level of nutrition. Economically, high levels of supplementation with expensive concentrates would not be appropriate but during periods of high demand such as late pregnancy and early lactation, supplementary feeding could be justified depending on the quality of the grazing. The increased level of nutrition would result in heavier kid birthweights and better lactation from the dam resulting in better kid growth. Improved maternal nutrition, during late pregnancy, will also improve the subsequent fibre yield of the progeny.

If the price of breeding stock for the fibre producing breeds is high, an increased standard of husbandry, including nutrition, can be justified as robust goats, with good growth rates and reproduction and low mortality, will give a better return from stock sales.

In the sort of more extensive systems appropriate to fibre goats the improvement they will effect on poorer pasture will be another valuable aspect. Quite a lot of research work has been carried out to show that goats can be instrumental in improving the production from other species, such as sheep, by selectively eating many of the plants and shrubs that inhibit grass growth in areas such as upland grazing where more conventional systems of improvement are difficult and expensive.[1] Interestingly, goats do not seem to compete with sheep very much in that they do not show any preference for grass or clover but spend most of their time browsing weeds and shrubs. This lack of competition means greater stocking densities are possible with mixed sheep and goats than would be the case with sheep alone.

Reproduction

Most aspects of reproduction and breeding of fibre goats are identical to the dairy breeds. It is the management of the goats during this period that may differ. In more extensive systems it is usual to run males with the females during the breeding season just as rams are run with ewes. If the pedigrees are important then a selected male may be run with a particular group of chosen females. If the timing of matings is required this can be achieved by using a ram harness, as described in Chapter 5, and the females will be marked as they are mated. A maximum of about forty females would be run with one male.

As milk is not required for sale it is usual to leave kids to be reared by their mothers and they would then be weaned at about 4–5 months of age. They would only be reared artificially in emergencies, if a female was unable to feed twins or if particularly fast growth rates were required. This latter point may be the case if the goats were being bred for sale. The system of artificial rearing would be the same as described in Chapter 5.

Health

Fibre goats are unlikely to be affected by any problems not seen in the dairy breeds. Obviously external parasites could have a more devastating effect by spoiling the fleece. Control of worms will be less of a problem because, as milk will not be sold, regular drenching is possible.

In typical British environments, where the ground will be relatively soft, Angora goats may have foot problems or at least need very regular hoof trimming. Angoras have evolved in relatively harsh, dry and rocky environments and in those conditions their hooves would not need trimming, whereas in our conditions they will require regular attention to avoid problems. This can be partially overcome by providing them with a pile of rock to jump about on. If rock is unavailable rough-cast building blocks are a good substitute.

HARVESTING THE FIBRE

Mohair

Angora goats are normally shorn twice a year, just before kidding and just before mating, the ideal length being 9–15 cm, which would normally be achieved in 5–6 months of growth. The preparation before shearing is most important as the clip or fleece will be the culmination of 6 months' work and perhaps years of selective breeding.

The goats should be on clean grazing prior to shearing and should not be run with other animals or any coloured goats. They must be shorn before natural shedding or cotting occurs. Cotting is when the fleece starts to lift away in one closely matted layer usually when a goat has been left too long before shearing.

The shearing premises must be scrupulously clean and if a sheep shearing facility is used it will be necessary to vacuum clean and wash away all traces of sheep wool. Dogs should not be allowed in the vicinity of the shearing pens as a single dog hair could result in a bale or swage of fleeces being downgraded.

Ideally a shearing facility should comprise one or two catching pens in which the goats will be held prior to shearing preferably with slatted or expanded mesh floors so that the fleeces do not become soiled or contaminated. The shearing area should have a smooth wooden floor, possibly constructed from a large sheet of plywood. This will be more comfortable for the goat and will be easy to sweep clean after each goat is shorn.

A sheep-shearing machine is suitable for goats as long as the speed can be varied. The fineness and smaller amount of grease in the fleece tends to make the clippers run hot. This is unpleasant for the shearer and can also damage the fleece. It is important that the speed is reduced about 25 per cent below sheep shearing speed and that a special goat comb of 17–20 teeth should be used on the cutter head. Regular oiling of the cutter head will also help keep it cool during shearing.

The method of shearing is very similar to that for sheep as can be seen in Figure 10.2. The biggest difference is in the preparations necessary to avoid spoiling the fleece. Because mohair is so much more valuable than sheep wool, and because certain contaminants or impurities can so adversely affect the finished product, it is necessary to go to great lengths to avoid contamination.

The care of the goat prior to shearing and the construction of the

Figure 10.2   Shearing a male Angora goat at Pound Farm in Devon.

shearing facility have already been mentioned. There is, however, a lot more that can be done to avoid penalties when it comes to selling the fibre. It is most important to avoid contamination with synthetic fibres and this means all plastic binder twine and plastic 'hessian' type sacking, etc. must be kept well away from the shearing area.

The goats should be shorn in order of fineness with the kids first and the adult males last. Pure bred goats should always be shorn before grade goats, as should white goats before coloured.

During the actual shearing it is most important not to make second cuts, i.e. cuts made when tidying up the goat by trimming the inevitable ridges left after the fleece has been cut away. These second cuts will obviously comprise short bits of hair which, if found when the fleece is being inspected, will result in it being downgraded. If the owner would like the goat to be tidied up this should be done after the main fleece has been taken away.

The entire fleece should be gathered up and taken to a clean, raised wooden work surface where it should be skirted. This is the process of picking out all the bits that would cause it to be downgraded. These would include bad stains, short cuts, cotted pieces and any areas with a lot of kemp, and they would be bagged separately.

Commercial mills pay the highest prices for white fibre because it can be dyed any colour. Some hand-spinners may prefer coloured fleeces to enable them to produce natural coloured garments.

The fleeces should be weighed and then packed. It should be noted that, as mohair readily absorbs moisture, fleece weights will vary according to atmospheric conditions. Fleeces of similar quality or from goats of similar age can be packed together. It is important that the fleeces are dry when packed or otherwise they will rot. Each fleece, along with its identification number, should be packed between paper in a special fleece sack or swage. This will be made from natural fibre and never plastic. As already mentioned bits of plastic getting into mohair is very serious and because of the problems it would cause during processing carries the heaviest penalty if found. Once packed in this way the bags can be tied or sewn up ready for sale or storage until required for sale or processing.

Cashmere

The harvesting of cashmere is not quite as straightforward as it is for mohair and as far as the United Kingdom is concerned different methods are still under investigation. The problem is that cashmere is produced as an insulating layer by goats in relatively cold environments and in Britain this is during the winter. Cashmere is shed or moulted when the weather begins to improve and temperatures rise, i.e. in spring. It can be seen, therefore, that it is necessary to harvest the cashmere at a time when yields will be optimal but before it is shed.

In the parts of the world where most cashmere is produced, such as the People's Republic of China, the cashmere is harvested by combing. It is unlikely, particularly if yields are low, that such a labour intensive method would be economic in Britain. The alternative is to clip the goats, especially as the larger processors now have the machinery to separate the down fibre from the primary guard hairs. The problem with this method is that goats in cold environments do not thrive if shorn of their protective coat. Housing for the time necessary for the goats to become acclimatised to their shorn condition would probably be uneconomic in terms of capital and variable costs.

One possibility that is being investigated is to fit the goats with a 'body stocking' that will prevent the cashmere from being lost when it is shed. It can then be collected when the stocking is removed. The ideal stocking has not yet been designed but this technique is showing promise.

Another possibility that is currently being studied is the use of the hormone melatonin.[2] This was mentioned in Chapter 5 in connection with the control of the breeding season. It seems that it may also play a part in the control of hair moulting or shedding. It is possible, therefore, that it could be used to control shedding so that it occurs at a more favourable time with the additional benefit that the fibre would be longer and, therefore, yields would be greater.

Once collected it is likely that British-produced cashmere would be packed in a similar way to mohair so that individual clips could be identified for the purpose of recording performance. This is most important, as great improvements, particularly in yield, are necessary if cashmere production on a large scale is to become viable.

Cashgora

This fibre, especially as it is the product of cross-breeding, is very variable. At its best a goat may produce enough to justify shearing twice a year like pure-bred Angoras. Others may produce much less once a year. Whatever the amount, like cashmere, there is good reason for individual clip identification if improvements in yield are sought.

To be able to breed selectively towards better fibre yields it is necessary to know the performance of each goat, in terms of the quality and quantity of the fibre it produces.

MARKETING GOAT FIBRE

Mohair

One of the attractions of farming Angora goats in Britain is that 40 per cent of the world mohair production, some 8,600 tonnes, is imported into this country for processing. There is, therefore, a potential ready market for British producers assuming that they are able to produce the quality of mohair required.

Mohair is purchased according to its quality and buyers require it to be classified into lines so that they are able to bid for bales or lots of known quality. It is at this stage that the preparation at shearing pays off because each bag or sack of around 50 kg will be classified according to the quality of the mohair it contains and any impurities will result in a considerable reduction in price.

The main classes for mohair are:

Superfine kid	23 microns or less
Kid	23–26 microns
Young goat	26–30 microns
Adult	30–34 microns
Strong adult	over 34 microns

If large quantities of mohair are produced these classes may be sub-divided even further mainly according to length. It can be seen that to be able to make best use of this type of marketing it is necessary to have enough mohair to be able to separate it into 50 kg lots according to its quality or class.

The British Angora Goat Society operates a fibre pool where members can send their fleeces in small lots and where they will be put with other fleeces of comparable quality to make up a bale. It is then possible to attract buyers from the major Bradford mills to bid for this mohair and the best possible price can be obtained. If the mohair is not classified the price achieved will reflect the poorest quality of fleece within the sample or bale.

Certification trade mark applied for

MOHAIR

Figure 10.3 The International Mohair Association's mohair trademark.

The amount of mohair being produced in Britain is increasing and the BAGS, at the time of writing, is taking steps to develop the fibre pool initiative along more commercial lines in order to cope with the increases in production. At present the fibre pool operates in May and November to coincide with twice-yearly shearing. Cashgora fibre is also handled by the fibre pool.

In addition to the possibilities of selling mohair to the mills and other processors in Bradford other companies have been set up to take mohair from production through to finished garments in the form of hand-produced designer knitwear. These companies will also buy mohair from producers.

There is considerable incentive for a producer or group of producers to add value to their own mohair by spinning and dying or even by knitting garments or weaving it into cloth. The right kind of garments can increase the value of the original raw fleece by as much as 20 times. This kind of cottage industry approach may be most appropriate in areas that attract large numbers of tourists. If garment production is sufficiently well organised marketing should be no problem, as there is a great deal of interest in quality designer knitwear throughout the country.

One problem that will arise will be the question of the coarser adult mohair which may not be suitable for fashion use. When mohair is sold to commercial mills this coarser fibre will be used for carpeting and other industrial uses.

Cashmere

The options for marketing cashmere are similar to those for mohair except there are not nearly so many processors or manufacturers working with it. Virtually all the commercial cashmere processing and manufacture in Britain is carried out by one company using approximately 1000 tonnes per annum. They estimate they could immediately use the cashmere produced by a million goats.

The same opportunities exist for adding value to cashmere for producers who are able to organise spinning and knitting. Cashmere is used almost solely for the fashion trade and there will not be the problem of finding uses for the coarser fibre as is the case with mohair.

REFERENCES

1. Russel, A. J. F., and Lippert, Margaret (1987), 'Goats and hill pasture management', in *Scottish Cashmere*, eds Russel, A. J. F. and Maxwell, T. J. (The Scottish Cashmere Producers' Association, Edinburgh).
2. Betteridge, K., Welch, R. A. S., Devantiar, B. D., Pomroy, W. F. and Hapwood, K. R. (1987), 'Melatonin-induced out-of-season growth of cashmere fibre in goats', in *Proceedings of the 4th International Conference on Goats*, (Brasilia).

Chapter 11

Goat's Meat

The market for goat's meat is expanding and more and more producers are beginning to find outlets for it.[1] This has been brought about by the greater interest in speciality foods and also by the increase in large goat herds which has overcome some earlier problems of finding regular supplies of goat carcasses.

BREEDS FOR MEAT

As outlets for goat's meat develop it will be appropriate to consider which breeds or crosses may give the best carcasses. Although more goats are kept throughout the world for meat than any other product there are almost no improved meat breeds. In view of this it is interesting to note that most of the improved dairy breeds found in Britain produce kids that, with a favourable rearing regime, can show quite impressive weight gains and will yield a high percentage of usable lean meat when slaughtered.

Of the important dairy breeds the Anglo-Nubian shows the most promise for meat production. Some can grow to impressive mature weights and they are slightly less boney than the Swiss breeds. Another breed which may be useful, particularly as a meat sire to put to dairy females, is the Angora. This breed is much more 'sheep shaped' than the dairy breeds as it has a broad back, short legs and a better general conformation. An added attraction of this breed is the marketable fibre that the cross-bred kids will produce by the time they reach a reasonable slaughter weight. As the numbers of Angoras throughout the country increase, males of not very good fibre quality but with good conformation can be purchased for a few hundred pounds and would be useful as meat sires.

One experimental study did show that the performance of Angora crosses on an artificial rearing system did not compare with that for British Saanen kids in terms of growth rate and feed conversion.[2] One explanation was that the Angora cross-bred kids did not seem to settle in the rearing environment and it would be useful if these differences could be researched further. Some years earlier the opposite was seen from a group rearing system when Angora crosses achieved the best weight gain of all. The Boer goat from South Africa is the only truly improved meat goat. These impressive animals look like extremely muscular Anglo-Nubians, with mature males weighing up to 150 kg. It is quite probable that the interest and knowledge in animal breeding in Britain will soon result in improved meat goats if the continuing expansion of goat's meat markets justifies such a development.

REARING FOR MEAT

As discussed in Chapter 5 kids do not normally grow very quickly when out to grass and, therefore, if market outlets justified fast growth, an artificial rearing regime would give the best results. This would be necessary anyway if milk was being sold from the herd. It would be particularly important to reduce rearing costs as much as possible and, therefore, the kids would be weaned early, i.e. 6–8 weeks, as the milk feeding part of the regime is the most costly. More research is needed to ascertain what would be the most economical post-weaning regime. The choice is from virtually *ad lib* concentrate through to some form of by-product forage with little or no concentrate.

For live weights up to 35 kg even British Saanen kids can grow at a rate of 220 g/d thus reaching the target weight by about 19–20 weeks. To achieve this they would eat a minimum of about 100 kg of concentrate feed.

An alternative to this expensive rapid growth system would be to have a relatively low concentrate, high forage regime. This could incorporate a grass or lucerne cube which would increase the protein in and the digestability of the feed without too large an increase in cost. Reducing the concentrate input will reduce weight gain but it could be that slower, less expensive growth would be more economic in some circumstances.

Grass reared kids left with their mothers might be successful on particularly clean pasture if the grass was good and the mothers were producing plenty of milk. A sheep creep system could be used that would allow the kids, but not their mothers, access to concentrate feed and possibly fresh pasture. If heavy worm burdens were thought to be a risk regular drenching would be necessary. This system would be most appropriate for fibre-producing herds.

As with other species entire males grow more quickly than those that are castrated. Nevertheless if male kids are to be slaughtered older than 4 months of age, particularly if this means during the mating season, it is likely that the meat will be tainted. Certainly when promoting kid meat it is not worth risking male taint even if the alternative is a slightly longer growth period.

CARCASS CHARACTERISTICS

For those used to assessing the quality of lamb carcasses, goat kid carcasses look rather scrawny with the exception of a few such as older Angora cross-breds. The main depot for fat in the goat is in the abdomen and most of this will be dressed out when the carcass is butchered. This lack of carcass fat is seen by some as an advantage, with current market trends, but it is sometimes difficult to convince traditional butchers of the potential value of a lean goat carcass.

It can clearly be seen from Table 11.1 that goat carcasses are more lean but also have more bone than lamb carcasses. What is not apparent from this data is the fact that, with goat carcasses, the yield of meat and the

Table 11.1   A comparison of lamb and kid carcasses

	Muscle	Bone	Subcutaneous fat	Intermuscular fat	Kidney fat
21 kg lamb carcass	55	12	16	17	4.1
20 kg kid carcass (dairy breed)	55.9	15.4	6.7	11.3	8.1
20.5 kg kid carcass (Angora × British Saanen)	56	14.6	12.5	17.0	4.6

proportion of fat do not change very much with increasing age or weight. It has been shown that Saanens, for example, can be taken to 0.65 per cent of their mature weight with little effect on the joint weight distribution or the total meat yield.[3]

Although little work has been done to evaluate different breeds it can be seen from Table 11.1 that the Angora cross carcass falls between the dairy kid and the lamb in that it has less bone and more fat than the Saanen. This is not surprising if one looks at Figure 11.1 which shows the rear view of a British Saanen and an Angora cross kid, both approximately 6 months of age. It would be fair to describe the Angora cross as sheep shaped.

MARKETING GOAT'S MEAT

When the subject of goat's meat is discussed among goat-keepers more often than not many will say there is no market for it and that the prices offered do not make it worthwhile to rear unwanted kids. In more recent years, however, it is possible to find some goat-keepers who have found or

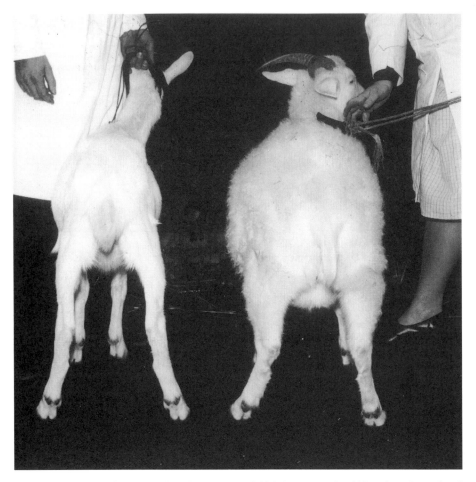

Figure 11.1   British Saanen and an Angora cross British Saanen wether kid at about 6 months of age (courtesy of the Food Research Institute, Reading).

developed outlets where prices are achieved which can make rearing profitable.

One of the main problems is being able to produce a regular and consistent supply. If a chain of stores or even an individual butcher is going to try a new line such as kid's meat they will want to be guaranteed enough to make promotion worthwhile.

A few years ago when the majority of goats were kept in small numbers this was more of a problem. Now that there are a significant number of herds of several hundred goats, regular supplies should be possible, particularly if breeding throughout the year can be achieved.

Figure 11.2 Goat meat exhibited by the Goat Producers' Association at the Royal Smithfield Show (courtesy of Dr J. M. Wilkinson).

A number of farmers could get together to form a meat-producing co-operative with kids being reared centrally on one farm. It is possible that grants from the Co-operative Development Board may be available for such an enterprise. Whatever approach is adopted it will be important to organise rearing so that a regular supply of carcasses of about the same weight can be guaranteed.

Persuading butchers or other retail outlets to sell kid's meat may take some ingenuity on the part of the farmer. A good move would be to produce a fact sheet describing the qualities and characteristics of goat's meat. A recipe sheet could also be offered. A carcass could be offered free of charge for the butcher to try. If this is done it should be of the very best quality. Never offer a carcass from an entire male unless it is very young, i.e. less than 4 months of age.

Experts in the butchery trade suggest that kid's meat is distinctive enough to be promoted in its own right as a speciality meat selling to that sector of the market that is prepared to pay top prices for top quality food that offers something different. This would possibly mean a different approach to butchering the carcass rather than following traditional lamb cuts. The new boned cuts used now for some lamb would seem particularly suitable for kid meat.

Senior staff at the Smithfield School of Food Technology have concluded, 'The meat should be positioned at the top end of the meat market
as an attractive alternative food, and sold at a price which reflects its
value.'[4]

One problem that has been anticipated by some is the lack of a suitable
name. The name goat conjures up, to some, images of old, tough, hardy
animals with a propensity for eating anything and by their very nature
giving the impression of being tough if not totally inedible. Some people
seem to find problems in accepting the idea of eating kid, as it seems to
suggest cannibalism! From the number of exhibitions where the Goat
Producers' Association has displayed and sometimes sold kid's meat it
would appear that there is not much of a problem. It may be that the
average meat-buying resident of the United Kingdom has become more
adventurous and open minded about what he/she will eat and this may
mean, in the end, that the name 'kid' is quite acceptable.

In France goat's meat is called chevon or if from very young animals
chevrette. In the USA and many Spanish-speaking countries it is called
cabrita. Some people even call it venison.

Scientists at the former Meat Research Institute at Bristol (now the
Food Research Institute Bristol Laboratory) found that goat meat had a
high ultimate pH in comparison with other meats, leading to a dark red
colour and a high water-holding capacity. These are positive attributes for
a meat to be used in processed products.[5] From this it may be concluded
that there are opportunities for producing sausage meat and pâté from
goat's meat.

With the marketing of any products it all comes down to mostly the
same key points and this is equally true with meat. It is important to be
able to supply meat of consistently good quality, the carcass size and type
should not be too variable, and whatever orders or quantities are agreed,
these should be reliably met. The goat-farming industry has now reached
the point where it is possible to satisfy all of these points.

REFERENCES

1. Mowlem, A. (1987), 'The development of a goat meat industry in Britain',
   British Goat Society Monthly Journal, Vol. 80.
2. Mowlem, A., Treacher, T. T., Wilde, Renee and Nash, S. J. (1987), 'The
   production of meat and fiber from a dairy goat herd', *Proceedings of the 4th
   International Conference on Goats* (Brasilia).
3. Butler-Hogg, B. W. and Mowlem, A. (1985), 'Carcass quality in British
   Saanen goats', *Proceedings of British Society of Animal Production*, in *Animal
   Production*, Vol. 40.

4. Wilkinson, J. M. and Barley, Jasmine B. (1986), 'Preparing cuts of goat meat at the Smithfield College', notes produced for the Goat Producers' Association's Exhibition at the Royal Smithfield Show.

5. Wood, J. D. (1984), 'Composition and eating quality of goats meat', in *Developments in Goat Production 1984*, eds Wilkinson, J. M. and Mowlem, A. (The Goat Producers' Association, Reading).

Chapter 12

Making a Profit

Although the majority of goats in Britain are kept for pleasure rather than profit, this book is intended for the growing number of prospective and actual goat farmers whose main interest in goats will be as a means of earning an income.

It is of course impossible to keep goats, whatever one's motives, without deriving a great deal of pleasure from them and this will be true for all goat farmers. Nevertheless goat farmers would not enjoy their goats for long if they did not make a profit. By profit we mean more money coming out of the enterprise than is going in and it is acknowledged that some farmers will not be exactly sure how much money is going in to their enterprise each year. It must also be borne in mind that the capital tied up in a goat-farming business would also earn an income if invested and to be successful a business would need to earn more than this.

The monetary values given in this chapter will probably be the first thing in the book to become out of date. However, the principles described will remain the same and the various comparisons between the different enterprises will remain more or less valid.

Before going into details about the financial aspects of the various types of goat production it is necessary to give the meaning of a number of terms used to avoid the confusion of different interpretations of the information given:

1. The *Gross Margin* of an enterprise is the output or monetary return from the products sold less the variable costs incurred in producing them.
2. The *Variable Costs* in the case of goats might include the cost of feed, veterinary and medical, processing and marketing.
3. The *Fixed Costs* of an enterprise will include those costs not included as variable costs and, in spite of the description, may not be fixed for long. Fixed costs normally include labour, machinery and equipment, building work, rent and depreciation.
4. *Finance Charges* or the cost of borrowed money would vary according to the circumstances of the farmer. Someone starting farming after redundancy from some other industry may have enough capital, in the form of a redundancy payment, to make borrowing unnecessary. However,

the value of that capital must be considered because of its earning potential if invested, as mentioned earlier.

5. The *Profit* will be the financial return achieved after payment of the fixed and variable costs of the enterprise plus a share of the fixed costs of other parts of the farm if there is more than one enterprise.

The parts all these components play in the cost and profit equation are demonstrated more clearly in Table 12.1.

It is normal for any enterprise to take several years to earn a greater return from capital than what can be achieved through normal investment channels.

It is very important when calculating projected profits or losses of any new enterprise to be pessimistic rather than optimistic. Things rarely go as well as is hoped. A good example of this is the milk yield from goats. The tremendous potential goats have for milk production has been discussed in Chapter 2 but it is unwise to base one's financial planning on record breaking yields. It takes goats a long time to settle in a new environment

Table 12.1 The relationship of the various cost (input) and income (output) components for costing a goat enterprise

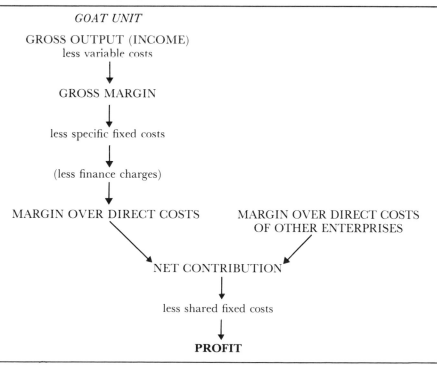

GOAT UNIT

GROSS OUTPUT (INCOME)
    less variable costs
    ↓
GROSS MARGIN
    ↓
less specific fixed costs
    ↓
(less finance charges)
    ↓
MARGIN OVER DIRECT COSTS   MARGIN OVER DIRECT COSTS
                                 OF OTHER ENTERPRISES
            ↘   ↙
            NET CONTRIBUTION
                ↓
        less shared fixed costs
                ↓
            PROFIT

Table 12.2 Variable cost incurred per goat in a 100-milker unit

	£
Concentrates	60
Forage	35
Straw	10
Vet. & med.	10
Herd replacer rearing costs	5
Chemicals	3
Heating/electricity	5
Transport	5
Total	133

Table 12.3 Fixed cost items that would apply to most goat enterprises

	£
Machinery: maintenance and repairs	2
Depreciation: herd	20
equipment (£10,000 over 5 years)	20
building costs	5
Insurance	4
	51

and even though 1000 l of milk per annum may be a realistic target it is unlikely the herd average will be much above 600 l in the first year.

These costs would apply to average milking goats producing between 600–1000 l of milk per annum but every farm will be different and it may be necessary to modify these figures to suit particular circumstances.

The feed costs may be slightly greater if all goats were averaging 1000 l per annum or they may be less if home grown forage or cereals were being used. It would be difficult to reduce many of the other items. In Table 12.4 a cost for milk packaging has not been included as it is intended to show the simplest situation of milk being sold in bulk to a processor who provides returnable containers.

The next thing to consider is the fixed costs and for this example it is assumed that no paid labour is used and that the farmer's income will come from the net profit from the enterprise.

Other fixed costs that might be added, depending on the circumstances on the farm, would be contractors, accountants and other professional fees, wages and cost of herd replacements.

Table 12.4 A comparison of gross margins from milk production

	Low	Average	High	High cow
Milk yield (l)	600	800	1200	5750
Milk price (p/l)	30	30	31	18
Milk value (£)	180	240	372	1035
Progeny value less depreciation	4.15	4.15	4.15	35
Output	184.15	244.15	376.15	676
Variable costs	133	143	153	394
Gross margin	51.15	101.15	233.15	676
Gross margin per hectare	409.50	609.20	1865.20	1284.40

Capital costs must also be taken into account as most enterprises will need start-up capital. Apart from the cost of capital, in terms of interest repayments if borrowed, the value of capital as potential investment must also be considered. There would be little point in developing an enterprise where the return achieved from the capital was not as great as could be achieved from ordinary forms of investment.

INCOME OR OUTPUT

The income derived from milking goats will obviously depend on the price paid for the milk and the amount produced per goat. In the example used in Table 12.4 it is assumed that no packing costs are incurred and the milk is being sold in bulk. A reasonable price for bulk milk sold to a processor would be 30p/l.

	£
600 l @ 30p	180
Surplus kid sold for meat (less cost and depreciation)	4.15
	184.15

PROFIT

From all this information it is now possible to calculate if the 100-goat enterprise would make a profit in the first year of production. The calculation would be:

Variable costs	100 × £133	= £13,300
plus		
Fixed costs	100 × £51	= £5,100
Output	100 × £184.15	= £18,415
Output less costs	£18,415 − 18,400	= £15 (profit before capital costs)

GROSS MARGINS

Because of the differences in capital requirement according to individual circumstances it is usual, with farm enterprises, to calculate the gross margin, particularly when comparing one enterprise with another. The figures in the example would give gross of £184.15 − 133 = £51.15. With a stocking rate of 8 goats plus followers to the hectare this would give a gross margin per hectare before depreciation of £409.20.

If we look at the same herd in Year 2 (second column in Table 12.4) an improvement can be seen resulting from the improved yields as the goats settle down and after some selection has taken place. Most commercial dairy goat farmers would be aiming to produce a herd average of 1000 l of milk per goat per year. If one refers to the yields discussed in Chapter 2 it can be assumed that this is well within the capabilities of British breeds.

If the price achieved for the milk increased by 1p this would of course increase the profitability of the 100-goat 1000-l enterprise by £1000 per year.

As with most goat products there is scope for adding value to the milk by processing it. The simplest product would be pasteurised milk for liquid consumption and the most technically involved would be hard cheese. To produce pasteurised milk for retail sales, capital expenditure would be necessary and an extra variable cost would be incurred in the form of packaging.

If we take the herd we have already considered we can modify the figures to give details of the profitability of such an enterprise. To pasteurise and package 1000 l of milk would cost approximately £120. If purchased new, the pasteurising and packaging equipment would cost approximately £12,000.

The value of the packaged milk, as pints, would be about 44p/l wholesale or 62p/l retail. Using the same calculation for the gross margin this would give, for the 800 l/goat herd, a gross margin per goat of £191.15 or £1529.20 ha.

For probably no additional costs or expenses, but with more technical expertise, even more value could be added by making cheese or yoghourt. One thousand litres of milk would make approximately 650 l of yoghourt with a wholesale value of £850 or 125 kg of hard cheese with a wholesale value of about £600.

If products are retailed from the farm an extra cost would be incurred in advertising and promotion. Eventually, if a reputation for good quality products is established, this cost may be reduced.

PROFIT FROM FIBRE

The cost and income equation for fibre production is quite different from the one for milk. With fibre goats the variable costs will be much less, mainly due to a greater reliance on grazing and less supplementation with concentrate feeds. Less capital equipment will be required but the goats themselves, at the present time, are much more expensive than dairy breeds. This does mean a major proportion of the output (income) will come from the sale of breeding stock.

Mohair

The example in Table 12.5 is for a herd of 100 pure-bred female Angoras which are valued at the current average price of £1000 each and therefore producing progeny with a reasonably high value for sale as future breeding stock.

As with the milking goats there is some scope for altering these costs. Feed costs can vary quite a lot especially if some feedstuffs are grown on the farm.

The fixed costs for an Angora herd are also likely to be less than for

Table 12.5 Variable costs incurred per goat in a 100-doe Angora herd

	£
Concentrates	15
Forage	8
Straw	2
Vet. & Med.	8
Shearing	4
Miscellaneous and transport	2
	39

Table 12.6 Fixed costs that would apply to most Angora herds

	£
Machinery	2
Depreciation	125
Insurance	5
	132

Table 12.7 Output or income from one female Angora goat

	£
Adult clip	30
Kids clip	16
Sale of kids	253
Sale of culls	3
	302
Cost of rearing replacements	− 1.6
	300.4

a milking herd except that the higher value will result in a greater depreciation.

Again, like the dairy enterprise, there may be other fixed costs such as contractors, accountants, wages and the cost of herd replacements.

The capital costs are made up from just the cost of the goats and the cost of fencing and shelters. The goats will cost £100,000, the fencing approximately £1,000 and the shelters approximately £1,500.

The annual income from the herd will come from the mohair clipped from the does and their progeny at about 6 months of age, the sale of female kids as breeding stock and the sale of wether males. Depending on their quality and origins the adult females could produce up to 8 kg of mohair per annum. For the purposes of the example in Table 12.7, 5 kg is used. Also the price obtained for mohair can vary year by year. An average of £6 for adult mohair and £8 for kid, which has been as much as £18 in recent years, is used in the example.

For the example in Table 12.7 a weaning rate of only 100 per cent has been used. It looks from recent experience as though it would be realistic to assume a 150 per cent weaning and with selection and good husbandry this could rise to the 200 per cent seen in dairy breeds. Obviously if the price of raw mohair remains at the high levels seen in recent years this also will make a considerable difference.

Once again as with the dairy goat examples it is now possible to work out a figure for net income exclusive of labour costs as it is assumed that this will be an owner worked enterprise:

Output 100 × £371 − (100 × variable costs £39 + fixed costs £132) = Net income before capital repayment £20,000

For exactly the same reasons as for the dairy goat example it is useful to calculate the gross margin for mohair production and in this case it is probably most appropriate to compare them with high yielding lowland sheep.

In the examples given in Table 12.8 an attempt has been made to show a range of likely situations that could occur as the market price for mohair and the value of breeding stock fluctuates. In recent years the price for mohair obtained by the producer has ranged from £7–18 for superfine kid

Table 12.8 A comparison of gross margins from mohair production

	A	B	C	D	E	F Sheep
Fleece yield (kg)	3	5	8	8	3	3
Fleece value £/kg	4	6	6	6	3	1.5
Progeny's 1st fleece (kg)	1.5	2	3.5	3.5	1.5	–
Kid's fleece value (£/kg)	6	8	12	12	6	
Female progeny value (£)	300	600	800	150	100	70
Sale of kids/lambs (£)*	138	253	329	82	63	31.9
Sale of culls (£)**	3	3	5	5	3	6
Ewe premium	–	–	–	–	–	3
Cost of rearing replacements (£)	1.6	1.6	1.8	1.8	1.6	5.0
Output	160.4	300.4	422.4	175.2	82.2	63.5
Variable costs (£)						
Concentrates		15				8
Forage (inc. bought in)		11				9
Vet. & med.		8				4
Shearing		4				1
Misc. & transport		2				1.5
Total variable costs		40				23.5
Gross margin per female	120.4	260.4	388.4	135.4	42.2	40
Gross margin per forage hectare	1204	2604	3884	1354	422	360

* The figure for the sale of kids/lambs is based on 100% weaning with 76% of the females' kids and 56% of the lambs sold for breeding, rest for replacers and wether sales.
** The cull figure is based on a replacement rate of 12% for the Angoras and 22% for the sheep.

and £3–8 for adult. This rise and fall of commodity prices is something most farmers have not had to worry about up to now as most commodities enjoy some degree of price support. This situation may change and if so more farmers will be affected by the state of the market.

The most pessimistic example in the gross margin calculations, column E, interestingly shows a margin figure that is still comparable with the one for a high yielding lowland ewe in column F.

Cashmere

Cashmere production, in Britain, is still very much in the developmental stage with improvements still being made in the productivity of the goats. When calculating the profitability of cashmere production it is difficult at this stage to quantify the value of the goats on unimproved upland pasture, a particular feature of this type of enterprise. As this type of goat is better suited to upland conditions it is appropriate to compare them with upland sheep.

Even greater reliance would be made of natural grazing with cashmere goats with just a minimum of feed supplementation during late pregnancy and early lactation.

The proportion of forage to concentrate costs is high because of the problems of growing forage in upland areas. The cost of harvesting the cashmere is high because, at present, labour-intensive methods are used. Again the fixed costs would vary according to the value of the goats which would also affect depreciation and insurance. The fixed costs may also include labour, contractors and administration.

Capital costs would include purchase of the goats and these may vary from less than £100 for feral base females to around £600 for imported improved cashmere-type goats, probably from Australasia. Another significant cost would be fencing, bearing in mind the problems of fencing

Table 12.9   Typical variable costs incurred per goat in a cashmere herd of 100

	£
Concentrates	8
Forage	6
Vet. & med.	3
Fibre harvesting	2
Misc. & transport	2
	21

Table 12.10   A comparison of gross margins from cashmere production

	A	B	C	D Upland sheep
Fibre yield (kg)	0.1	0.3	0.5	3.0
Fibre value (£/kg)	60	60	60	1
Female progeny value	50	100	300	50
Sale of kids/lambs (£)*	26.5	55.5	131	36.2
Sale of culls (£) **	3	3	4	5
Cost of rearing replacements (£)	1.4	1.4	1.5	2
Subsidy	–	–	–	8
Output	34.1	75.1	163.5	50.2
Variable costs (£)				
Concentrates	8	8	9	5
Forage (inc. bought in)	6	7	7	6
Vet. & med.	3	4	4	3
Harvesting fibre	1	2	2	1
Misc. & transport	2	2	2	2
Total variable costs	20	23	24	17
Gross margin per female	14.1	52.1	139.5	31.2
Stocking rate (hectare)	13	13	13	9
Gross margin per forage hectare	183.3	677.3	1813.5	289.8

* Except for replacers all females sold for breeding and all males for meat.
** Replacement/culling rate 12% goats 22% sheep.

some upland terrain. It could cost as much as £2/m depending on the site and the quality of the fencing used. It is likely that some cashmere farmers in less-favoured areas will be eligible for grants towards such items as fencing. Because fixed and variable costs can be so variable it is, once again, appropriate to look at the gross margin calculation to assess the potential of this type of enterprise.

As with Angoras several examples are given illustrating the current situation with various ways of starting such an enterprise particularly with respect to the quality and productivity of the goats.

It is reasonable to expect that good stock will be in demand for some years as cashmere production develops. As with Angoras this will mean that the selling of breeding stock will produce a significant proportion of the income from a cashmere herd.

PROFIT FROM GOAT'S MEAT

The trade in goat's meat is by no means developed throughout Britain. Problems of finding or developing outlets may make it impossible for some farmers to consider meat production as a viable addition to their goat enterprise. However, some farmers have developed lucrative outlets and some ideas for this are discussed in Chapter 11.

As with the other enterprises discussed so far, a number of examples of the possible returns for goat's meat production will be given, each based on figures from actual situations.

The variable costs will differ quite a lot between enterprises depending on such factors as natural or artificial rearing, type of milk replacer used, weaning age and costs of other feeds.

Fixed costs would include labour, machinery and building maintenance and machinery and buildings depreciation.

Specific capital costs would be small. Even if the kids were being bought as orphans they would only represent a few hundred pounds unless a very large kid-rearing enterprise was planned.

This example is typical of the example given in Table 12.1 at the beginning of this chapter showing how cost and returns from various enterprises on a farm interrelate to give a profit figure for the farm as a whole. It is likely that kid-rearing would be run in conjunction with other enterprises such as milk or fibre production.

As with all farm enterprises there are many factors concerning kid rearing that will be different from farm to farm. Once again, the gross margin calculation would be an appropriate way of looking at the viability of the various options.

It can be seen from Table 12.12 that rearing kids can only pay if the rearing costs are reduced to a minimum or if some lucrative outlet can be found, both are possible. Although rearing the kids on their mothers is the lowest cost system it may not be the most realistic as the only kids reared

Table 12.11 Individual variable costs incurred for artificially rearing kids for meat

	£
Milk replacer and concentrates	28.75
Hay	1.80
Bedding	1.00
Vet. & med.	0.50
Misc. & transport	0.50
	32.55

Table 12.12   A comparison of gross margins for kids reared for meat on different systems

	High-class butcher	Livestock Market	Retail from farm
36 kg kid value (£)	40	22.5	45
Skin value	5	–	6
Less cost of kid	5.25	5.25	5.25
Output	39.75	17.25	45.75

Variable costs (£)			
		A B C	
Milk replacer and concentrates*		28.75 35.9 10.0	
Hay		1.8 2.5 –	
Misc.		2.0 2.0 2.0	
Total variable costs		32.55 40.4 12.0	

	A	B	C	A	B	C	A	B	C
Gross margin per kid	7.25	0.65	27.75	−15.3	−23.15	5.25	13.2	5.35	33.75

A = Kid weaned at 6 weeks, lamb-bar *ad lib* feeding.
B = Kid weaned at 8 weeks, lamb-bar *ad lib* feeding
C = Kid reared on mother, out to grass.
* The feed costs are based on commercially-available calf milk replacers and concentrate rations, not the more expensive products produced specifically for kids.

in this way, that would be available for meat, would be cross-bred fibre males. Other fibre goats would not be killed for meat and dairy goats would be artificially reared to release milk for sale.

Kids reared on grass grow more slowly and this may mean they could miss the best selling period. Also, because their growth is less predictable, it is more difficult to guarantee to sell kids of a particular weight at a specific time.

These figures also emphasise the need for good marketing to achieve a good return from meat kids. It should be remembered that fixed and capital costs and finance charges still have to be subtracted from these gross margin figures before a net profit can be calculated.

CAN GOAT FARMING BE PROFITABLE?

This chapter has shown a wide range of examples and possibilities where goats can be profitable. Nevertheless it can also be seen that there are many factors that influence the profitability of a goat enterprise.

In more orthodox agriculture, farmers have not yet had to face the reality of market-led production. The goat farmer is in the vanguard of this movement where profits can only be made if there are market outlets for a product within an economic distance from the farm. In the case of dairy products this may mean a nearby distribution warehouse for one of the large retail multiples.

One thing that can now often be seen in other farm enterprises is the principle of adding value to products. With goat's milk there is the potential for making cheese or yoghourt, and goat's fibre can be spun, dyed and knitted or woven into fashion clothing or cloth.

Goat enterprises do fail. Poor husbandry is rarely a reason but unsound finances or financial management is often the cause. From the examples given in this chapter it can be seen that some enterprises require a lot of expensive equipment. It is difficult to start such an enterprise without access to comparatively large sums of money and if borrowed these will incur large repayment charges.

Those who have kept goats as pets may be at a disadvantage when it comes to running a commercial goat enterprise. Such people often find it difficult to adjust. For example a commercial enterprise cannot afford to carry unproductive goats as pets and any that are poor producers or any that cause problems must be culled.

Some factors influencing the profitability of a goat enterprise are difficult to cost such as natural weed control. Regardless of any profit that may be made, most people who are involved with the husbandry of goats will agree that they are remarkable animals and that the pleasure they can give can sometimes go a long way towards compensating for the sometimes small financial return.

At its best goat farming can be at least as profitable as any other livestock farming and in some cases much more so. In this age of agricultural change goats offer farmers a viable alternative and an interesting challenge.

Appendices

1. FURTHER READING

Books

Devendra, C. and Burns, Marca (1970), *Goat Production in the Tropics* (Commonwealth Agricultural Bureau, Farnham Royal).

Devendra, C. and Mcleroy, G. B. (1982). *Goat and Sheep Production in the Tropics*, (Longman, London).

Dunn, P. (1982), *The Goatkeepers' Veterinary Book* (Farming Press, Ipswich).

Gall, C. (ed.) (1981), *Goat Production* (Academic Press, London).

Guss, S. B. (1977), *Management and Diseases of Dairy Goats* (Dairy Goat Publishing Corporation, Scottsdale, Arizona).

Holmes Pegler, H. S. (1927), *The Book of the Goat*, 9th ed (The Bazaar, Exchange and Mart Ltd, London).

La Jaouen, J.-C. (1987), *'The Fabrication of Farmshead Goat Cheese'*, *Cheesemakers' Journal* (Cashfield, Mass.).

Mackenzie, D. (1980), *Goat Husbandry*, 4th edn rev. and ed. J. Laing (Faber & Faber, London).

Morris, J. and Cave-Penny, T. (eds) (1987), *Goats for Fibre* (The National Angora Stud, Bodmin).

Ørskov, E. R. (1987), *The Feeding of Ruminants: Principles and Practice* (Chalcombe Publications, Marlow).

Quittet, E. (1975), *La Chevre* (La Maison Rustique, Paris).

Ryder, M. L. (1987), *Cashmere, Mohair and Other Luxury Animal Fibres for the Breeder and Spinner*. (White Rose II, Southampton).

Thiel, C. C. and Dodd, F. H. (eds) (1979), *Machine Milking, Technical Bulletin No. 1* (National Institute for Research in Dairying, Reading).

Westhuysen, J. M. vander., Wentzel, D. and Grobler, M. C. (1985), *Angora Goats and Mohair in South Africa* (South African Mohair Growers' Association).

Wilkinson, J. M. and Stark, Barbara A. (1987), *Commercial Goat Production* (BSP Professional Books, Oxford).

Reports and Conference Proceedings

Developments in Goat Production, 1984, 1985, 1986–87 (The Goat Producers' Association).

Fullerton-Smith, H. P. (1986), *The Study of Group Breeding Schemes* for Sheep and Their Application to Breeding Up Cashmere and Angora Goats (Nuffield Farming Scholarship Trust, Olney, Bucks).

Grainger, Mary (1986), *The Marketing of Goat Meat* (National Farmers Union, Marketing Division, Stamford).

Islay and Jura Goat Society, (1985), *Goat Husbandry Survey Report*, (Isle of Islay, Argyll).

Proceedings of the International Symposium on Nutrition and Systems of Goat Feeding, Tours, France (1981), (ITOVIC, Paris).

Proceedings of the Third International Conference on Goat Production and Disease, Tucson, Arizona, (1982) (Dairy Goat Journal Publishing Corporation, Scottsdale, Arizona).

Proceedings of the Fourth International Conference on Goats, Brasilia (1987) (EMBRAPA, Brasilia).

Proceedings of the Third Symposium on the Machine Milking of Small Ruminants, Valladolid, Spain, (1983) (MAPA, Madrid).

Proceeding of a Conference on Mohair or Cashmere Production – New Farm Enterprise, (1987) (Royal Agricultural Society of England).

Proceedings of a Seminar on The Angora Alternative, (1987) (The British Angora Goat Society, Malvern).

Scott, Susan A., (1984), *Commercial Goat Dairying*, (MAFF/ADAS).

Steele, M. A. (1984), *'Aspects of Marketing Goat Meat'*, M.Sc. thesis, Centre for Tropical Medicine, University of Edinburgh.

Thornhill, S., (1984) 'Commercial dairy goat farming in the UK', B.Sc. Thesis, Department of Agriculture, University of Reading.

Vincent, R. J., (1983), *Commercial Goat Farming*, Report on a visit to Australia, New Zealand and the USA in June to August 1982 to study the opportunities for UK and Dairy Goat Farming (Nuffield Farming Scholarship Trust, Olney, Bucks).

Webb, K. R. (1985), 'Buildings for the UK goat industry: a basis for design', B.Sc. thesis, Department of Agriculture, University of Reading.

Journals and Periodicals

British Angora Goat Society Newsletter, British Angora Goat Society, Three Counties Showground, Malvern, Worcestershire.

Fibre News, Official Journal of the Mohair Producers' Association of New Zealand and the Cashmere Producers of New Zealand, PO Box 641 Whangarei, New Zealand.

Goat Producers' Association Newsletter, Goat Producers' Association, N.A.C., Stoneleigh, Warwickshire.

Goat Veterinary Society Journal, The Goat Veterinary Society, Chalk Streets, Rettendon Common, Essex.

Home Farm, Broad Leys Publishing Co., Widdington, Saffron Walden, Essex.

La Chèvre, Monthly Journal, ITOVIC, 149 rue de Bercy, Paris.

Monthly Journal of the British Goat Society, British Goat Society, Bovey Tracey, Devon.

Smallholder, Monthly Magazine, High Street, Stoke Ferry, King's Lynn, Norfolk.

Small Ruminant Research, Journal of the International Goat Association, Elsevier Publications, Amsterdam.

The Ark, Journal of the Rare Breeds Survival Trust, N.A.C. Stoneleigh, Warwickshire.

2. Useful Addresses

The British Angora Goat Society
Three Counties Agricultural Society
Malvern
Worcestershire

The British Goat Society
34–6 Fore Street
Bovey Tracey
Devon

The Goat Producers Association
National Agricultural Centre
Stoneleigh
Kenilworth
Warwickshire

The Goat Veterinary Society
The Limes
Chalk Street
Rettendon Common
Essex

The Royal Agricultural Society of England
National Agricultural Centre
Stoneleigh
Kenilworth
Warwickshire

The Rare Breeds Survival Trust
National Agricultural Centre
Stoneleigh
Kenilworth
Warwickshire

Ministry of Agriculture, Fisheries and Food
Whitehall Place
London
SW1A 2HH

Agricultural Development and Advisory Service
Great Westminster House
Horseferry Road
London
SW1P 2AE

National Farmers Union
Agriculture House
25–31 Knightsbridge
London
SW1X 7NJ

The Scottish Cashmere Producers Association
c/o SAOS Ltd
Claremont House
12–19 Claremont Crescent
Edinburgh
EH7 4JW

3. Laws and Acts Concerning the Goat Farmer

Copies of all these can be obtained from Her Majesty's Stationery Office.

Diseases of Animals Act 1950
Concerns control of spread of disease particularly in the notifiable diseases (see Chapter 7).

Food Hygiene (general) Regulations 1970 – enforced by Environmental Health Officers and applies to processing of food, e.g. butter, cheese and yoghurt.

Labelling of Food Regulations 1970.

Movement of Animals (Records) Order 1960. Defines records to be kept in 'Movements Book'.

Protection of Animals Act 1911.

Under the Food and Drugs Act 1970 (as amended) come the Cheese Regulations 1970 (as amended) and the Dried Milk Regulations 1965.

Veterinary Surgeons Act 1966
Defines procedures which may be performed by persons other than qualified veterinarians.

Index

FARMING PRESS BOOKS

Below is a sample of the agricultural and veterinary books published by Farming Press. For more information or for a free illustrated book list please contact:

Farming Press Books, 4 Friars Courtyard, 30–32 Princes Street, Ipswich IP1 1RJ, United Kingdom. Telephone (0473) 43011.

The Goatkeeper's Veterinary Book

Peter Dunn

How to diagnose goat ailments, how to treat them, when to send for the vet and how to prevent further outbreaks of disease. Emphasis on good husbandry. Second edition.

All about Goats

Lois Hetherington

An introduction to all aspects of goatkeeping, covering feeding, milking, kidding, housing, rearing, management, products and showing.

Alternative Farm Enterprises

Bill Slee

Essential reading for farmers in today's rapidly changing marketplace. Discusses in detail how the farmer can use a professional marketing approach to diversify with confidence.

Farm Woodland Management

John Blyth, Julian Evans, William E. S. Mutch, Caroline Sidwell

A compendium for farmers in which all aspects of trees on the farm are dealt with in detail – planning, planting, harvesting, accounting.

A Way of Life: Sheepdog Training, Handling and Trialling

H. Glyn Jones and Barbara Collins

A complete guide to sheepdog work, in which Glyn Jones' life with sheepdogs is presented as an integral part of his practical advice on training, handling, trialling and breeding.

Outdoor Pig Production

Keith Thornton

Keith Thornton analyses the potential for outdoor units in the light of current technological changes, sets the financial background and gives details of the day-to-day running of an outdoor unit.

On Being a Tenant Farmer: A Layman's Guide to the Landlord and Tenant System

Henry Fell

A route map through the intricacies of the landlord and tenant system, providing the background knowledge to determine when professional help is needed and where to get it.

A Veterinary Book for Dairy Farmers

Roger Blowey

A new, enlarged edition of this authoritative volume that deals with the full range of cattle disorders grouped broadly according to the age and development of the animal from the young calf to adulthood.

Farm Building Construction

Maurice Barnes and Clive Mander

Covers all practical farm building work. Details on blockwork, brickwork, timber flooring, roads, walls, etc. for new constructions or improvements.

Calf Rearing

Bill Thickett, Dan Mitchell and Bryan Hallows

A new edition of this volume that reflects modern practices as experienced in a wide variety of situations covering housed rearing of calves to twelve weeks.

Farming Press also publish three monthly magazines: *Dairy Farmer, Pig Farming* and *Arable Farming*. For a specimen copy of any of these magazines please contact Farming Press at the address above.